Construction Hazardous Materials Compliance Guide

Lead Detection, Abatement, and Inspection Procedures

R. Dodge Woodson

ELSEVIER

AMSTERDAM • BOSTON • HEIDELBERG • LONDON
NEW YORK • OXFORD • PARIS • SAN DIEGO
SAN FRANCISCO • SINGAPORE • SYDNEY • TOKYO
Butterworth-Heinemann is an imprint of Elsevier

Butterworth–Heinemann is an imprint of Elsevier
225 Wyman Street, Waltham, MA 02451, USA
The Boulevard, Langford Lane, Kidlington, Oxford, OX5 1GB, UK

Library of Congress Cataloging-in-Publication Data
Woodson, R. Dodge (Roger Dodge), 1955-
Construction hazardous materials guide : lead in remodeling, renovation,
  and repair projects / Roger Woodson.
    p. cm.
ISBN 978-0-12-415838-2
  1. Lead based paint—Removal—United States. 2. Lead based paint—Law
and legislation—United States. 3. Lead abatement—Law and legislation—United
States. I. Title.

TD196.L4W66 2012
363.17'91—dc23   2011052962

British Library Cataloguing-in-Publication Data
A catalogue record for this book is available from the British Library.

For information on all Butterworth–Heinemann publications
visit our website at *http://store.elsevier.com*

Printed in the United States
12 13 14 15 16   10 9 8 7 6 5 4 3 2 1

*This book is dedicated to Afton and Adam
for always being there for me.*

Contents

Contents

Introduction

If you are a contractor who works with older homes and buildings you may well encounter lead-based building materials. When this happens many situations can arise. Bidding a job that will require specialist treatment to contain or abate lead dust without including the work in your bid can result in a financial disaster. Should you disturb lead paint in any way your actions are likely to release dangerous lead-containing particles into the surrounding air. There are serious health risks associated with lead paint and related materials. Negligence on your part can result in major lawsuits. Lead paint is not a material to be taken lightly and any older building can contain some of this hazardous material.

Whether you are a remodeling contractor, a carpenter, a plumber, a heating mechanic, a roofer, a siding contractor, or a flooring contractor you can be at risk. In 2008, the Environmental Protection Agency released a new lead paint rule, "Renovation, Repair, and Painting Rule." This rule changed the playing field for contractors, requiring them to become familiar with it before working on older buildings.

The laws, regulations, and rules governing working with lead-based materials are set forth by both federal and state agencies. Any failure to comply with the requirements detailed by governing authorities can result in stiff fines and potential lawsuits. These requirements are extensive and can be complex.

R. Dodge Woodson is a career contractor who has been both a remodeling contractor and a plumbing contractor since 1979. In this book he provides invaluable insight and guidance for those who find themselves working with lead paint. There is a wealth of information contained within, ranging from Woodson's field experience to the many rules, regulations, and laws on both state and federal levels.

If you are someone who has any chance of encountering lead paint in your work, you need this invaluable guide to keep you informed and safe. Look at the Contents, then thumb through the pages. You will see quickly see that this is a comprehensive guide for all types of contractors. Do not let yourself be taken off guard. Read this book and prepare yourself for dealing with lead hazards in future job sites.

► ACKNOWLEDGMENTS

I want to thank the Environmental Protection Agency and the Occupational Safety and Health Administration for information used in the preparation of this book.

► ABOUT THE AUTHOR

R. Dodge Woodson is a career contractor with more than 30 years of experience. He has been a master plumber, builder, and remodeling contractor since 1979. The Woodson name is synonymous with professional reference books. The author has written numerous best-selling books over the years.

Lead Basics

Lead is common in older homes and buildings. Any building constructed before 1980 is a likely suspect for containing lead, since lead paint was used for many years before its health risks were fully understood. If young children eat flaking paint—a common problem—that contains lead, the health repercussions can be very serious. Structures with lead-based paint include schools, multifamily housing, single-family homes, and so on. Unfortunately, control, containment, and abatement can be extremely expensive.

Contractors who are bidding renovation and remodeling work must be particularly concerned with the risk of lead paint existing in a structure. Fairly simple tests can be conducted to indentify the presence of lead. This is always a good investment for any contractor before a firm price for a job is committed to.

Government regulations and housing regulations pertaining to lead are strict. Failure to comply with the rules and regulations can result in lawsuits and fines. Basically, contractors cannot afford to take a chance that lead is present when they are remodeling or renovating buildings. See Box 1.1 for some facts about lead.

Box 1.1 Facts about Lead

FACT: Lead exposure can harm young children and babies even before they are born.

FACT: Even children who seem healthy can have high levels of lead in their bodies.

FACT: You can get lead in your body by breathing or swallowing lead dust, or by eating soil or paint chips containing lead.

FACT: You have many options for reducing lead hazards. In most cases, lead-based paint that is in good condition is not a hazard.

FACT: Removing lead-based paint improperly can increase the danger to your family.

► HEALTH EFFECTS OF LEAD

Childhood lead poisoning is a major environmental health problem in the United States. See Box 1.2 for an explanation of why children are at higher risk. Everyone can get lead in their bodies if they do any of the following:

- Put their hands or other objects covered with lead dust in their mouths
- Eat paint chips or soil that contains lead
- Breathe in lead dust, especially during renovations that disturb painted surfaces

The danger is highest for children and the effects on them usually occur at lower blood-lead levels than for adults. For examples of the effects that lead can have on adults, see Box 1.3.

Children who have high levels of lead in their bodies can suffer from:

- Damage to the brain and nervous system
- Behavior and learning problems, such as hyperactivity
- Slowed growth
- Hearing problems
- Headaches

Box 1.2   Reasons Why Lead Is More Dangerous to Children

- Babies and young children often put their hands and other objects in their mouths; such objects can have lead dust on them.
- Children's growing bodies absorb more lead.
- Children's brains and nervous systems are more sensitive to the damaging effects of lead.
- Developing nervous systems can be affected adversely at blood-lead levels of less than $10\,\mu g/dL$.

Box 1.3   Adults and Lead

- Reproductive problems (in both men and women)
- High blood pressure and hypertension
- Nerve disorders
- Memory and concentration problems
- Muscle and joint pain

▶ LEAD REMEDIATION

Beginning April 22, 2010, federal law requires that contractors performing renovation, repair, and painting projects that disturb more than six square feet of paint in homes, childcare facilities, and schools built before 1978 must be certified and trained to follow specific work practices to prevent lead contamination. Contractors working with lead are required to be in a lead-safe certified company. Many houses and apartments built before 1978 have paint that contains lead (called lead-based paint). Lead from paint, chips, and dust can pose serious health hazards if not taken care of properly.

Federal law requires that individuals receive certain information before renting or buying pre-1978 housing. Landlords must disclose known information on lead-based paint and lead-based paint hazards before leases take effect. Leases must include a disclosure form about lead-based paint. Property sellers must disclose known information on lead-based paint and lead-based paint hazards before selling a house. Sales contracts must include a disclosure form about lead-based paint. Buyers have up to 10 days to check for lead hazards.

Lead is a big concern that contractors need to be aware of. The risk of taking lead containment lightly can be extremely expensive for contractors. Imagine the following scenario.

You have just signed a contract to paint a large, two-story house that was built in 1974. The existing paint is peeling and flaking on the siding and the interior paint work is extensive. Being a competitive contractor, you kept your bid as low as possible. Your profit margin is modest, but a profit is a profit.

The homeowner is having the house painted to help sell the house. Your crew arrives and starts sanding the exterior siding and filling cracks in the weathered wood. A different crew is prepping the interior walls, ceilings, and trim for new paint. Your crews are busy working when a home inspector arrives on the site. The homeowners called the inspector in to get advice on a hairline crack in the building's foundation. It isn't long before your phone is ringing with a panicked homeowner on the other end of the phone.

It seems that the inspector noticed the repair work your crews were doing and became concerned. As a trained, experienced inspector, he knew the house was likely to contain lead-based paint. He decides a risk assessment is in order. (See Table 1.1 for an example of available statistics for a risk assessment.) The inspector told the homeowners about his concern for the family, which included two young children, your crews, and neighboring houses that could be affected from blowing dust as the exterior siding was being sanded down to remove the existing paint.

The homeowner requested a lead test from the inspector and called you to stop the work until test results were available. This is terrible news for you. It disrupts your production schedule and could put you on the hook for legal problems and state and government fines. As a result, you stop your crews and start to look into the requirements of working with lead paint. You should have done this long before you put a bid on the job, but you didn't.

After going nuts worrying about the test results, you get the phone call that confirms the worst possible outcome. The paint inside and out is lead paint. You have already created several violations of lead regulations and the homeowners are speaking with an attorney.

TABLE 1.1 Percentage of Paint that Is Lead-Based for Use in Risk Assessment

COMPONENT TYPE	INTERIOR	EXTERIOR
Walls/Ceiling/Floor		
1960–1979	5	28
1940–1959	15	45
Before 1940	11	80
Metal Components[1]		
1960–1979	2	4
1940–1959	6	8
Before 1940	3	13
Nonmetal Components[2]		
1960–1979	4	15
1940–1959	9	39
Before 1940	47	78
Shelves/Others[3]		
1960–1979	0	—
1940–1959	7	—
Before 1940	68	—
Porches/Others[4]		
1960–1979	—	2
1940–1959	—	19
Before 1940	—	13

[1]Includes metal trim, window sills, molding, air/heat vents, radiators, soffit and fascia, columns, and railings
[2]Includes nonmetal trim, window sills, molding, doors, air/heat vents, soffit and fascia, columns, and railings
[3]Includes shelves, cabinets, fireplace, and closets of both metal and nonmetal
[4]Includes porches, balconies, and stairs of both metal and nonmetal
Source: HUD 1990b. These data are from a limited national survey and may not reflect the presence of lead in paint in a given dwelling or jurisdiction.

See how bad this could get? You need to know the facts about lead if there is any chance you will be working with it.

Where Lead Is Found

In general, the older a building the more likely it has lead-based paint. This is the type of lead that affects most contractors. But, lead can also be found in drinking water, soil, old toys, old furniture, some pottery that has been painted with lead paint, and even in some routine job situations. All of these are a potential concern, but contractors need to be the most concerned about lead paint.

The federal government banned lead-based paint from housing in 1978, though some states stopped its use even earlier. Lead can be found in homes in the city, country, or suburbs. It can be in apartments, single-family homes, and both private and public housing, in exterior and interior paintwork.

Lead can also be found in soil. It can come from the soil around a home or other building (though few contractors are affected by this) where it is picked up from exterior paint or from other sources such as past use of leaded gas in cars. Children playing in yards can ingest or inhale lead dust.

Even household dust can contain lead. The dust can pick up lead from deteriorating lead-based paint or from soil tracked into a home. The drinking water for a building could contain lead. A potential reason for this is the use of lead plumbing joints and lead-based solder. Today's plumbers must use lead-free solder, but that was not the case for many years.

A phone call to your local health department or water supplier can direct you to a testing source for lead content. You cannot see, smell, or taste lead, and boiling your water will not get rid of lead.

Buildings served by a water source that may contain lead should be tested. In the meantime, people under these conditions should do the following:

- Use only cold water for drinking and cooking.
- Run water for 15 to 30 seconds before drinking it, especially if the water has not been used for a few hours.

It is important for those working with lead to shower and change clothes before going home or to another location because it can be brought with you on hands or clothes. Work clothes should be laundered separately from the rest of the family's clothes.

Where Lead Is Likely to Be a Hazard

Lead from paint chips, which you can see, and lead dust, which you can't always see, can be serious hazards. Lead-based paint that is in good shape and has not been disturbed poses less of a problem. However, there is always a risk that the lead paint will be disturbed, especially by children chewing on it. Another major risk is the remodeling or renovation of a building. Hazards are highest under the following conditions:

- Peeling, chipping, chalking, or cracking lead-based paint is a hazard and needs immediate attention.
- Lead-based paint may also be a hazard when found on surfaces that children can chew or that get a lot of wear and tear. These areas include:
  - Windows and window sills
  - Doors and door frames
  - Stairs, railings, and banisters
  - Porches and fences

Lead dust can form when lead-based paint is dry scraped, dry sanded, or heated. Dust also forms when painted surfaces bump or rub together. Lead chips and dust can get on surfaces and objects that people touch. See Figures 1.1 and 1.2 for examples

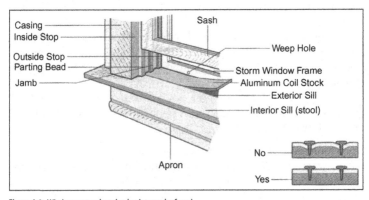

Figure 1.1 Window parts where lead paint may be found.

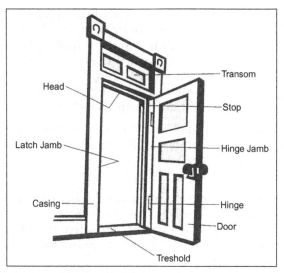

Figure 1.2 Door sections where lead paint may be found.

of likely areas for lead paint to be found. Settled lead dust can reenter the air when people vacuum, sweep, or walk through it. Buildings that use duct work to distribute air conditioning and heat can also spread lead dust.

Lead in soil can be a hazard when children play in bare soil or when people bring soil into the house on their shoes. The National Lead Information Center (NLIC) is a good resource for finding out how to test soil for lead.

▶ HOW TO DETECT LEAD

Just knowing that a building has lead-based paint may not indicate the level of hazard. You can check buildings for lead content in a couple of ways. A paint inspection can reveal the lead content of every different type of painted surface in a building. However, it won't tell you whether the paint is a hazard or how you should deal with it. A risk assessment will determine if there are any sources of serious lead exposure (such as peeling paint and lead dust). It will also tell you what actions to take to address these hazards.

It's wise to have qualified professionals do the work if lead risks must be addressed. There are standards in place for certifying lead-based paint professionals to ensure the work is done safely, reliably, and effectively. The National Lead Information Center (NLIC) compiles a list of contacts by area.

Trained professionals use a range of methods when checking a building for lead. Some of these methods include:

- Visual inspection of paint condition and location
- A portable x-ray fluorescence (XRF) machine
- Lab tests of paint samples
- Surface dust tests

Amateur test kits for lead are available, but studies suggest they are not always accurate. These tests should not be relied on before doing renovations or to assure safety. It's wise, and a good investment, to have a professional conduct the testing with specialized equipment.

Contractors may not want to pay for testing before they have secured a job, but there is a workaround. They can make their bid based on the assumption that no negative environmental test results will be determined once the job is awarded and before the work is started. The bid can be worded to allow for an increase in the bid price if professional testing determines that additional work is needed to deal with environmental risks.

What to Do to Protect Occupants While Awaiting Professional Help

There are times when lead contamination is expected or confirmed. Due to scheduling requirements, however, the occupants of a building may have to live with the risk for a while until the contractors can perform their services. If you have a customer with such a situation, there are a few things that can be done to minimize the effects of lead for a short period of time, including the items listed in Box 1.4. Figure 1.3 contains information about the toxicological effects of lead on both children and adults.

▶ LEARN THE REGULATIONS, RULES, AND LAWS

It is important to learn the regulations, rules, and laws pertaining to working with lead. There is plenty to learn. As a lead contractor, your work will fall under the regulation of the

Box 1.4   Steps to Take to Minimize Lead's Effects

- If you suspect that your house has lead hazards, take the following immediate steps to reduce your family's risk:
  - If you rent, notify your landlord of peeling or chipping paint.
  - Clean up paint chips immediately.
  - Clean floors, window frames, window sills, and other surfaces weekly. Use a mop, sponge, or paper towel with warm water and a general all-purpose cleaner or a cleaner made specifically for lead. Never mix ammonia with bleach products. Doing this can result in a very dangerous gas.
  - Thoroughly rinse sponges and mop heads after cleaning dirty or dusty areas.
  - Wash children's hands often, especially before they eat and before nap time and bed time.
  - Keep play areas clean. Wash bottles, pacifiers, toys, and stuffed animals regularly.
  - Keep children from chewing window sills or other painted surfaces.
  - Clean or remove shoes before entering your home to avoid tracking in lead from soil.
  - Make sure children eat healthy and nutritious meals as recommended by the National Dietary Guidelines. Children with good diets absorb less lead.
- Additional steps:
  - You can temporarily reduce lead hazards by taking actions such as repairing damaged painted surfaces and planting grass to cover soil with high lead levels. These actions are not permanent solutions and will need ongoing attention.
  - To permanently remove lead hazards, you must hire a certified *lead abatement contractor*. Abatement (or permanent hazard elimination) methods include removing, sealing, or enclosing lead-based paint with special materials. Just painting over the hazard with regular paint is not enough.
  - Always hire a person with special training for correcting lead problems. You need someone who knows how to do this work safely and has the proper equipment to clean up thoroughly. Certified contractors will employ qualified workers and follow strict safety rules set by their state or the federal government.
  - Contact the NLIC for help with locating certified contractors in your area and to see if financial assistance is available.

Occupational Safety and Health Administration (OSHA), the Environmental Protection Agency (EPA), and your state's rules and regulations. Compliance with all of these is necessary.

As we explore working with lead in the following pages, you will become familiar with the types of rulings to expect as a lead contractor.

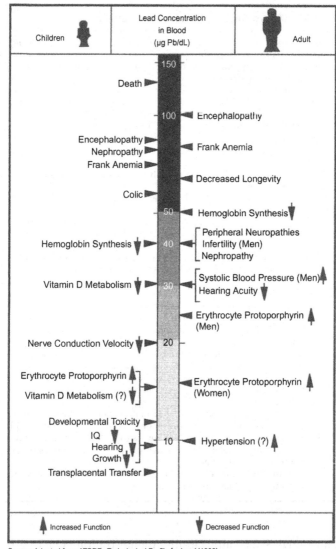

Source: Adapted from ATSDR, *Toxicological Profile for Lead* (1989).

Figure 1.3 Toxicological effects of lead on children and adults.

Lead Testing

2

The determination of lead by rapid methods has become a major concern in recent years since it is apparent that lead poisoning is still occurring when workers remove or otherwise disturb old lead-containing paint in the course of their work. Serious cases of lead poisoning continue to occur in children exposed to old flaking lead-containing paint or lead-contaminated soil. Abatement projects have become a major activity in recent years to remove lead-containing paint and other materials to reduce lead exposure to workers and the general public.

The term *lead-based paint* (LBP) has been defined in the Lead Exposure Reduction Act (October 29, 1992) as "paint or other surface coatings that contain lead in excess of 1.0 milligram per centimeter squared or 0.5% by weight" (Sections 10.7, 10.8). In the U.S. Department of Housing and Urban Development (HUD) Interim Guidelines (September 1990, revised May 1991, Section 10.9), the action level for LBP is a lead content of $1.0 \, mg/cm^2$ as measured by an XRF analyzer. When using chemical testing, the action level is either 0.5% by weight or $1.0 \, mg/cm^2$. *Lead-free paint*, on the other hand, is usually considered to be any dried coating containing no more than 0.06% (600 ppm) lead, essentially indicating no lead is added at any point in the manufacturing process.

OSHA currently does not stipulate a specific amount of surface lead that is not to be exceeded. The OSHA CSHO examines the industrial operation or situation for potential occupational exposure to lead as a surface contaminant, and evaluates protective measures to minimize exposures. Although OSHA does not have a numeric trigger for lead as a surface contaminant, concentrations in air are strictly regulated in occupational settings. For the purposes of this evaluation, the HUD guideline value of 0.5% by weight will be used as the trigger for evaluating lead test kit performance.

The HUD Interim Guidelines on Worker Protection includes discussions on worker training, enclosures, personal protective equipment, work practices, and personal hygiene. The guidelines are intended to provide, at a minimum, the level of protection of OSHA's General Industry Lead Standard. Both OSHA's General Industry and lead-in-construction regulations stipulate identification of potential exposure to lead and good housekeeping practices to minimize surface contamination.

Lead detection kits are useful as a quick check for screening areas for lead abatement. A positive response is evidence of the presence of lead or a positive interference. A negative response, however, is not conclusive evidence of the absence of lead. The test provides presumptive evidence for the presence of lead, not its absence. A more thorough determination may need to be performed by a quantitative laboratory analysis of any representative bulk material available to substantiate the absence of lead. Samples are analyzed for lead at OSHA's Salt Lake Technical Center (SLTC) using OSHA methods ID-121 with Atomic Absorption Spectroscopy (AAS), ID-125 G with Inductively Coupled Plasma (ICP), or ID-206 (Solders by ICP). If necessary, lower limits of detection for lead may be achieved using ICP Mass Spectrometer procedures.

The Lead Alert™ test kit now sold by Sensidyne® was originally developed by Frandon Enterprises, Inc. which is currently owned by Pace Environs, Inc. An extensive study reported by Pace Environs indicated that the methodology to evaluate paint samples met HUD performance standards for qualitative lead tests.

The Lead Check™ test kit now also sold by Sensidyne has been tested by the Food and Drug Administration (FDA) and the National Institute of Standards and Technology (NIST). The FDA reported that the swabs were able to detect the presence or absence of lead in 96.6% of the pieces tested. The NIST found that the swabs were able to detect lead in paint as well as or better than other methods. The product literature indicates that 1 to 2 µg of leachable lead can be detected.

▶ ADVANTAGES AND DISADVANTAGES OF LEAD DETECTOR TEST KITS

Lead detector kits are designed to offer a simple and immediate estimation as to whether or not the lead in paint or other solid materials is at a hazardous level and whether abatement measures are needed. See Box 2.1 for an example of the uses of a test kit. According to the manufacturers, in addition to paint, other materials such as dust, soil, ceramics, lead crystal, solder, foil, pewter, and other metals can be examined for lead. The kits are easy to use and require no analytical laboratory time. There can be interference with the testing. Box 2.2 shows examples of this.

Lead test kits are inexpensive. The pink color obtained on reaction of lead with the kit reagents is very distinct and easy to interpret. Tests appear to be specific for lead when a pink color is obtained. They do not give a positive reaction with several metals used in lead-free solders. The chemicals used are stated

---

Box 2.1 Use of Lead Test Kits

- Each kit is provided with test papers that contain lead to assure effectiveness in testing. Care should be used in handling these test media.
- The kits are not recommended for users who are color blind in the red/pink region of the color spectrum.
- The kits are not designed to detect lead in water.
- Possible interferences include barium, calcium sulfate in plaster, chromate in lead and zinc chromate, and red paint pigments. The kits give only a positive/negative response. For a more thorough lead determination, the samples must be sent to a certified laboratory, or another more expensive field procedure used.

Box 2.2   Potential Interferences for Lead Test Kits

Barium: Some paints contain barium sulfate (an extender) which will produce an orange color. If a pink color is observed over the orange, a positive result is indicated. However, the pink color may also be obscured, in which case the sample should be sent to a certified laboratory, or another field method used.

Gypsum, plaster dust, stucco: These contain calcium sulfate, a negative interferant. Sulfates present in these materials can interfere with color development.

Chromate: Some paints used on metal surfaces (on bridges, in tanks, etc.) contain lead chromate, which is extremely difficult to detect, especially for yellow and orange paints. Chromate is a negative interferant. When a negative result is obtained and there is reason to believe lead chromate is present, it should be confirmed by a certified laboratory or another field technique used. Zinc chromate may also interfere.

Red painted surfaces: Red pigment may interfere with a positive surfaces indication (pink color formation).

to be nontoxic. However, the usual precautions in handling any chemicals should be followed, and their use is not recommended for someone who is color blind in the color spectrum's red/pink region.

Using a lead test kit

The use of a lead test kit is not complicated. Like with most products, the user manuals for each kit must be consulted before use. OSHA performed tests with various types and brands of test kits. One of the kits they tested was the Lead Check test kit.

The contents of each kit include test swabs and ampoules of chemicals. The composition of the chemicals was not given. The instructions for use of the kit that OSHA used in creating its rules and regulations are as described next.

For all testing applications, the following steps are used to activate each swab:

1. *Crush*: With the swab tip pointing up, squeeze and crush points marked "A" and "B" located on the barrel of the swab.
2. *Shake and squeeze*: With the swab tip pointing down, shake twice and squeeze gently until the yellow liquid appears on the swab tip. The swab is now activated for testing.
3. *Rub*: While squeezing gently, rub the swab tip on the test area for 30 seconds.

If the swab turns pink, the test is positive and lead is present. If the swab indicates no color change, the test is negative and no leachable lead has been detected. All negative results must be confirmed using the test confirmation card. If the swab does not turn pink after testing, rub the swab tip on one of the unused test dots on the card. If a pink or red color appears on either the dot or swab, a negative result is substantiated. If the test dot and/or swab does not turn pink or red, the test was invalid and must be repeated with a new swab. All testing, including any necessary confirmation card testing, must be completed within two minutes.

> **Caution**: The test dots on the Test Confirmation Card contain minute amounts of lead. Do not touch these areas and be sure to wash your hands after each use.

Procedure for examining painted wood or metal surfaces

1. Clean and remove all dust and dirt from the area to be tested.
2. With a clean knife or scraper, cut a small notch at a diagonal to expose all painted layers down to the bare surface. Lead may be present in any layer of paint.
3. Activate a swab.
4. Rub the activated swab in the exposed cross-section for 30 s. If any layers contain lead, a positive result will occur and the swab or surface will turn pink.

Note: With certain paints, lead is difficult to extract, and it may take longer for positive results to develop. Examine the test surface 30 to 60 minutes after the test has been performed before assuming a test result.

Note: *Bleeding* may occur when testing surfaces that are painted red. Moisten cotton-tipped applicators with a few drops of distilled white vinegar. Rub the moistened cotton on the red surface. If a red color appears on the cotton, Lead Check swabs cannot be used. Send a sample of the material to a certified laboratory or use another type of field procedure for further testing.

Procedure for examining lead-containing ceramics or glass

1. Activate the swab.
   (a) When testing ceramics, rub the swab over all patterns with different colored glazes and any cracks or chipped areas. If lead is leaching from the surface, the swab will turn pink.
   (b) When testing glass, rub the swab over the surface. If lead is leaching from the surface, the swab will turn pink.
   (c) When testing lead crystal, rub the swab on the inside surface of the glass or other vessel. If lead is leaching from the surface, the swab will turn pink.

Procedure for examining solder (plumbing, food cans)

1. Wipe off the solder joint with a paper towel or cloth.
2. Using an emery board or sandpaper, lightly score the surface to be tested.
3. Activate the swab.
4. Squeeze one drop of yellow liquid from the swab onto the solder surface.

5. Touch the swab to the wet solder surface and rub gently for only 10 s or less.
6. The swab will turn pink if the solder contains greater than 2% lead.

Procedures for examining lead in dust

1. Activate the swab.
2. Rub the activated swab in the dust for 30 s. If the dust contains lead, the swab will turn pink.

Note: Avoid rubbing the swab into large amounts of "dirt dust" as this will obscure the color test results on the swab.

Another test kit that OSHA used was Lead Alert All in One™. The indicating solution contains rhodizonate ion which reacts with lead to give a pink color; the composition of the leaching solution was not given. For recovering paint down to the base material from painted objects, a kit containing a coring tool and cleaning brush is available for use with the Lead Alert kits. The initial preparation and testing of the indicating solution takes about 10 minutes. Most tests take about 1 to 3 minutes to perform. According to OSHA, the following are the use instructions for this kit.

Procedure for activating the indicating solution

1. Remove the red cap from the indicating solution bottle.
2. Carefully remove the dropper insert by rolling/twisting it off the bottle.
3. Open the foil package and place the indicating tablet into the bottle of solution.
4. Replace the dropper insert and the red cap. Tighten the cap and shake the bottle for 1 min.
5. Allow the bottle to stand for 5 min, then shake it again until the solution turns yellow. The tablet will not completely dissolve; however, this is normal. The indicating solution is now ready for use. (For reference purposes, write the date on the bottle when the reagent is activated.)
6. When testing has been interrupted for more than 15 min, shake the indicating solution vigorously again for 5 to 10 s before resuming testing. When the testing has been interrupted for several hours or days, shake the indicating

solution for 30 s and perform a positive control test before resuming testing.

Procedure for carrying out the positive control test to verify the effectiveness of the testing system

1. Place one drop of the leaching solution on the Positive Control Strip in the center of one of the numbered circles.
2. Let it sit for 10 s, then add one drop of the activated indicating solution.
3. A pinkish (rose) stain will appear in the circle and indicates the test kit is working properly. If the rose stain does not appear, the activity of the indicating solution has expired.

Caution: For testing purposes, the areas inside the circles on the positive control strip contain minute amounts of lead. Do not touch these areas and be sure to wash your hands after each use. After testing and when dry, put the positive control strip back into its pouch.

Procedure for carrying out a total lead test on paint

1. Remove a fresh adhesive-backed collection paper and fold it in half. Apply the paper close to the area to be tested. Using the coring tool, cut down into the surface. Scrape the paint from inside the circle and place it on the paper. **Be sure to remove all layers.** Thoroughly clean the coring tool. Transfer the paint from the paper to a plastic vial. Grind up the paint with a plastic rod for about 10 s.
2. Add three drops of leaching solution to the vial and grind the contents for another 10 s. Let the vial sit for 20 s.
3. Add three drops of indicating solution to the tip of a fresh sample collector. Touch the surface of the liquid in the plastic vial with the tip of the collector.
4. A pink color on the collector tip indicates lead in amounts greater than the limit of 0.5% by weight of lead.

Note: When testing red painted surfaces it is possible that the red paint may "bleed out" into the test surface on the sample collector. To test for bleeding out, look at the liquid above the solid in the vial in Step 4. If the fluid is red, the sample should be sent to a certified analytical laboratory for further testing or another type of field procedure should be used.

Procedure for carrying out a surface lead test on paint (top layer only)

1. Apply two drops of leaching solution to the tip of a sample collector.
2. Rub the sample collector tip on the surface to be tested for 10 to 15 s.
3. Add two drops of indicating solution to the sample collector tip.
4. A pink color on the collector tip indicates lead in amounts greater than the limit of 0.5% by weight of lead for the top layer of paint.

Procedure for carrying out the sanding test

This test should be used when only the top one or two layers of paint are to be sanded or otherwise disturbed in preparation for painting; if more than two layers are to be sanded or otherwise disturbed, use the total lead test described previously.

1. Use a fresh abrasive strip to rub lightly over the painted surface. An area of about 0.5 in^2 is sufficient. Avoid contact with or breathing of dust.
2. Apply two drops of leaching solution to a sample collector tip and rub it over the sanded area.
3. Add one more drop of leaching solution to the paint dust now on the collector.
4. Leave for 30 s. Add two drops of indicating solution to the same area on the collector tip.
5. A pink color on the collector tip indicates lead in amounts probably greater than the limit of 0.5% by weight of lead for the layer(s) of paint affected.

Note: Bleeding may occur when testing surfaces are painted red. During the sanding test on red paint, examine the surface of the sample collector after the leaching solution and paint dust have been added to the tip. If the color does not spread, the red pigment will not interfere with the test. If it does spread, send a sample of the material to be tested to a certified analytical laboratory or use another type of field procedure.

Procedure for determining lead in paint, metal, and dust particles

1. Apply two drops of leaching solution to a sample collector tip.
2. Apply a very small amount of fine particles of the target material (ground paint chips, paint dust, house dust, etc.) to the tip of the collector.

3. Apply one or two more drops of leaching solution to the particles on the tip. Wait 30 s.
4. Apply two drops of indicating solution to the collector tip; watch for a color change.
5. A pink color on the collector tip indicates lead in amounts probably greater than the limit of 0.5% by weight of lead.

Procedure for plumbing pipes, joints, and fixtures

1. Locate an area where water pipes are exposed. Determine if any soldered joints are present.
2. Sand the pipe and/or solder joint lightly with a fresh abrasive strip.
3. Apply two drops of leaching solution to the tip of a sample collector.
4. Rub the tip of the collector on the surface to be tested for 10 to 15 s.
5. Add two drops of indicating solution to the collector tip; watch for a color change.
6. A pink color on the collector tip indicates lead in amounts probably greater than the limit of 0.5% by weight of lead.

▶ CERTIFIED TESTING

Immediate test kits are good, but they should not be depended on for remodeling and renovation projects. The simple kits can give you a quick read on a job, but they are not considered to be certified testing. As a contractor you should only depend on the services of qualified, certified testers and laboratories for your results. This is the only way to ensure that you will not suffer repercussions down the road.

Preparing to Begin a Job

Preparing to begin a job can start with testing, but this is not all that is needed. Contractors have to work within the confines of many regulations. These will be explained in detail as we go along. Before we get into that area of knowledge let's consider some of the first steps for a job.

There is likely to be some type of work order or contract in place for the work you will do. Once you have your authorization to work, you will need to give notification of the work you will be doing. See Figure 2.1 for an example of this type of form. It is helpful to include a design of the scheduled work order in

⚡EPA

U. S. ENVIRONMENTAL PROTECTION AGENCY

NOTIFICATION

OF LEAD-BASED PAINT ABATEMENT ACTIVITIES

Important: A representative of the certified firm may complete this sample form or a similar form when notifying EPA. Consult the *Instructions for Notifying EPA Commencement of Lead-Based Paint Abatement Activities* when preparing abatement notification. **Please type or print responses in black or blue ink only.**

A. Type of Notification   Please indicate the type of notification.

☐ Original   ☐ Updated   ☐ Cancellation

B. Emergency Notification ☐   No   ☐   Yes, if yes include documentation showing evidence of an EBL determination or a copy of the Federal/State/Tribal/Local emergency abatement order.

C. Activity Start and End Dates   Specify the dates you will begin and end lead-based paint activity.

If necessary, estimate end date using   Start date: _____   End date: _____
your best professional judgment.   Month/Day/Year   Month/Day/Year

D. Description of Activity   This section relates to the building where abatement work will be performed.

Type of Building: ☐ Single Family Dwelling   ☐ Multi-Family Dwelling   ☐ Child-Occupied Facility
Property name (if applicable): _____

Property Address including apartment and/or unit number(s):

_____
Street Address   City   State   Zip Code

Square footage/acreage to be abated: _____

Please write a brief description of abatement project to be performed. (Enclose additional paper if necessary)

E. Firm Information
Name: _____   Firm's Certification Number: _____

Address: _____
                          Street Address   City   State   Zip Code
Phone Number: _____

F. Certified Supervisor's Information
Name: _____
EPA Certification Number: _____ (Check here ☐ if working under interim certification
and enter the identification number from your course completion certificate in this space)

G. Firm Affirmation   Please note that this form is incomplete without a signature.

I hereby attest and affirm that the information included on this notification form is true and accurate to the best of my belief and knowledge. I acknowledge that any approval authorized pursuant to this notification will be subject to revocation if issuance was based on incorrect or inadequate information that materially affected the decision to issue the approval.

Name: _____   Title: _____

Signature: _____   Date Signed: _____

For information on EPA and other lead programs, see the web site:
http://www.epa.gov/lead/

Figure 2.1 Notification form.

your files. After your files are up to date with beginning paperwork, you are ready to proceed to the site.

Risk assessment is a big factor to consider before establishing a work plan. This type of information is explained later, but we will touch on it here. There are also some protection steps to take into consideration.

Risk assessment takes on many forms. One is shown in Table 2.1. It has to do with the assessment of risk once a paint inspection, and possibly testing, is done. You will also want to consider the different types of work you will be doing. Each type can involve different rules and regulations. See Tables 2.2 and 2.3 for examples of different types of work.

Before work is started worksites must be prepared. What does this require? It depends on the work that will be done. There are different levels of risk for various types of work. Are you doing interior or exterior work? This will make a difference in your prep work. Recommendations on preparing worksites can be found in Tables 2.4 and 2.5. It is also worth noting that you will be required to put safety procedures into play. These will be dealt with in depth later, but as an example you will need to

TABLE 2.1 Comparison of Risk Assessment and Paint Inspection

ANALYSIS, CONTENT, OR USE	RISK ASSESSMENT	PAINT INSPECTIONS
Paint	Deteriorated paint *only*	Surface-by-surface
Dust	Yes	Optional
Soil	Yes*	Optional
Water	Optional	Optional
Air	No	No
Maintenance status	Optional	No
Management plan	Optional	No
Status of any current child lead poisoning cases	If information is available	If information is available
Review of previous paint testing	Yes	Yes
Typical applications	1. Interim controls 2. Building nearing the end of expected life 3. Sale of property/ turnover 4. Insurance (documentation of lead-safe status)	1. Abatement 2. Renovation work 3. Weatherization 4. Sale of property/ turnover 5. Remodeling/ Repainting
Final report	Lead hazard control plan or certification of lead-based paint compliance	Lead concentrations for each surface tested

*If local experience indicates that soil lead levels are all very low, repeated soil sampling is not necessary.

TABLE 2.2 Selected Renovation Jobs and Work Practices

	CONTAINMENT	RELOCATION	RECOMMENDED PRACTICES	CLEANUP
Demolition	Use plastic sheeting to prevent airborne dust migration Interior Worksite Prep. Level 4; Exterior Worksite Prep. Level 3	No residents in dwelling during any work	Wet surfaces, use covered containers to move debris; best subcontracted to abatement contractor, or a demolition contractor certified for abatement	HEPA vacuum, wet mop, and HEPA vacuum
Repainting	Floors and ground covered with 6-mil plastic Interior Worksite Prep. Level 4; Exterior Worksite Prep. Level 3	No entry into work area during interior work	Wet scrape, wet sanding, HEPA-filtered vacuum power tools	Daily cleanup with HEPA vacuum, wet wash, HEPA vacuum
Floor Sanding	Full containment of rooms, negative air recommended if leaded dust hazard identified	No entry into work area during work	Sanding lead-containing floors should be completed by abatement contractor, or other contractor certified for abatement	HEPA vacuum of entire house may be needed
Plaster Repairs	Localized containment for walls, entire room for ceiling Usually Interior Worksite Prep. Level 1 or 2 for small jobs	No entry into work area	Wet prior to removing	HEPA final cleanup
Window Replacement	Localized containment around each opening	No occupancy during removal and initial cleaning and sealing	Seal interior with plastic Remove window from exterior if possible	HEPA vacuum all areas with replaced windows
Carpet Removal	Do dust sampling to determine contamination level. Usually Interior Worksite Prep. Level 3 or 4	No occupancy during removal and initial cleaning	Carefully remove and package carpet and pad in 6-mil plastic with taped seams. Wet down carpet before removal or disturbance	HEPA vacuum floor after carpet bagged and prior to removal

TABLE 2.3 Window Treatment or Replacement Worksite Preparation

APPROPRIATE APPLICATIONS	ANY WINDOW TREATMENT OR REPLACEMENT
Resident Location	Remain inside dwelling but outside work area until project has been completed. Alternatively, can leave until all work has been completed. Resident must have access to lead-safe entry/egress pathway.
Time Limit per Dwelling	None
Containment and Barrier System	One layer of plastic sheeting on ground or floor extending 5 feet beyond perimeter of window being treated/replaced. Two layers of plastic taped to interior wall if working on window from outside; if working from the inside, tape two layers of plastic to exterior wall. If working from inside, implement a minimum Interior Worksite Preparation Level 2. Children cannot be present in an interior room where plastic sheeting is located due to suffocation hazard. Do not anchor ladder feet on top of plastic (puncture the plastic to anchor ladders securely to ground). For all other exterior plastic surfaces, protect plastic with boards to prevent puncture from falling debris, nails, etc. (if necessary). Secure plastic to side of building with tape or other anchoring system (no gaps between plastic and building). Weigh all plastic sheets down with two-by-fours or similar objects. All windows in dwelling should be kept closed. All windows in adjacent dwellings that are closer than 20 feet to the work area should be kept closed.
Signs	Post warning signs on the building and at a 20-foot perimeter around building (or less if distance to next building or sidewalk is less than 20 feet). If window is to be removed from inside, no exterior sign is necessary.

Cont'd

23

TABLE 2.3 Window Treatment or Replacement Worksite Preparation—cont'd

APPROPRIATE APPLICATIONS	ANY WINDOW TREATMENT OR REPLACEMENT
Security	Erect temporary fencing or barrier tape at a 20-foot perimeter around building (or less if distance to next building or sidewalk is less than 20 feet). Use a locked dumpster, covered truck, or locked room to store debris before disposal.
Weather	Do not conduct work if wind speeds are greater than 20 miles per hour. Work must stop and cleanup must occur before rain begins, or work should proceed from the inside only.
Playground Equipment, Toys, Sandbox	Removed from work area and adjacent areas. Remove all items to a 20-foot distance from dwelling. Large, unmovable items can be sealed with taped plastic sheeting.
Cleaning	If working from inside, HEPA vacuum, wet wash, and HEPA vacuum all interior surfaces within 10 feet of work area in all directions. If working from the exterior, no cleaning of the interior is needed, unless the containment is breached. Similarly, no cleaning is needed on the exterior if all work is done on the interior and the containment is not breached. If containment is breached, then cleaning on both sides of the window should be performed. No debris or plastic should be left out overnight if work is not completed. All debris must be kept in a secure area until final disposal.

TABLE 2.4 Interior Worksite Preparation Levels (not including windows)

DESCRIPTION	LEVEL 1	LEVEL 2	LEVEL 3	LEVEL 4
Typical Applications (hazard controls)	Dust removal and any abatement or interim control method disturbing no more than 2 square feet of painted surface per room.	Any interim control or abatement method disturbing between 2 and 10 square feet of painted surface per room.	Same as Level 2	Any interim control or abatement method disturbing more than 10 square feet per room.
Time Limit per Dwelling	One work day	One work day	Five work days	None
Resident Location	Inside dwelling, but outside work area. Resident must have lead-safe passage to bathroom, at least one living area, and entry/egress pathways. Alternatively, resident can leave the dwelling during the work day.	Same as Level 1.	Outside the dwelling; but can return in evening after day's work and cleanup are completed. Resident must have safe passage to bathroom, at least one living area, and entry/egress pathways upon return. Alternatively, resident can leave until all work is completed.	Outside the dwelling for duration of project; cannot return until clearance has been achieved.

Cont'd

TABLE 2.4 Interior Worksite Preparation Levels (not including windows)—cont'd

DESCRIPTION	LEVEL 1	LEVEL 2	LEVEL 3	LEVEL 4
Containment and Barrier System	Single layer of plastic sheeting on floor extending 5 feet beyond the perimeter of the treated area in all directions. No plastic sheeting on doorways is required, but a low physical barrier (furniture, wood planking) to prevent inadvertent access by resident is recommended. Children should not have access to plastic sheeting (suffocation hazard).	Two layers of plastic on entire floor. Plastic sheet with primitive airlock flap on all doorways. Doors secured from inside the work area need not be sealed. Children should not have access to plastic sheeting (suffocation hazard).	Two layers of plastic on entire floor. Plastic sheet with primitive airlock flap on all doorways. Doors secured from inside the work area need not be sealed. Overnight barrier should be locked or firmly secured. Children should not have access to plastic sheeting (suffocation hazard).	Two layers of plastic on entire floor. If entire unit is being treated, cleaned, and cleared, individual room doorways need not be sealed. If only a few rooms are being treated, seal all doorways with primitive airlock flap to avoid cleaning entire dwelling. Doors secured from inside the work area need not be sealed.
Warning Signs	Required at entry to room but not on building (unless exterior work is also under way).	Same as Level 1.	Posted at main and secondary entries, since resident will not be present to answer the door.	Posted at building exterior near main and secondary entryways.

Ventilation System	Dwelling ventilation system turned off, but vents need not be sealed with plastic if they are more than 5 feet away from the surface being treated. Negative pressure zones (with "negative air" machines) are not required, unless large supplies of fresh air must be admitted into the work area to control exposures to other hazardous substances (for example, solvent vapors).	Turned off and all vents in room sealed with plastic. Negative pressure zones (with "negative air" machines) are not required, unless large supplies of fresh air must be admitted into the work area to control exposure to other hazardous substances (e.g., solvent vapors).	Same as Level 2
Furniture	Left in place uncovered if furniture is more than 5 feet from working surface. If within 5 feet, furniture should be sealed with a single layer of plastic or moved for paint treatment. No covering is required for dust removal.	Removed from work area. Large items that cannot be moved can be sealed with a single layer of plastic sheeting and left in work area.	Same as Level 2

Cont'd

TABLE 2.4 Interior Worksite Preparation Levels (not including windows)—cont'd

DESCRIPTION	LEVEL 1	LEVEL 2	LEVEL 3	LEVEL 4
Cleanup	HEPA vacuum, wet wash, and HEPA vacuum all surfaces and floors extending 5 feet in all directions from the treated surface. For dust removal work, a HEPA vacuum and wet wash cycle is adequate (i.e., no second pass with a HEPA vacuum). Also wet wash and HEPA vacuum floor in adjacent area(s) used as pathway to work area. Do not store debris inside overnight; transfer to a locked secure area at the end of each day.	HEPA vacuum, wet wash, and HEPA vacuum all surfaces in room. Also wet wash and HEPA vacuum floor in adjacent area(s) used as pathway to work area. Do not store debris inside dwelling overnight; use a secure locked area.	Remove top layer of plastic from floor and discard. Keep bottom layer of plastic on floor for use on the next day. HEPA vacuum, wet wash, and HEPA vacuum all surfaces in room. Also wet wash and HEPA vacuum floor in adjacent area(s) used as pathway to work area. Do not store debris inside dwelling overnight; use a secure locked area.	Full HEPA vacuum, wet wash, and HEPA vacuum cycle.
Dust Sampling	Clearance only	Clearance only	One sample collected outside work area every few jobs plus clearance.	Clearance only

Note: Primitive air locks are constructed using two sheets of plastic. The first one is taped on the top, the floor, and two sides of doorway. Next, cut a slit about 6 feet high down the middle of the plastic; do not cut the slit all the way down to the floor. Tape the second sheet of plastic across the top of the door only, so that it acts as a flap. The flap should open *into* the work area.

TABLE 2.5 Exterior Worksite Preparation Levels (not including windows)

DESCRIPTION	LEVEL 1	LEVEL 2	LEVEL 3
Typical Applications	Any interim control or abatement method disturbing less than 10 square feet of exterior painted surface per dwelling. Also includes soil control work.	Any interim control or abatement method disturbing 10 to 50 square feet of exterior painted surface per dwelling. Also includes soil control work.	Any interim control or abatement method disturbing more than 50 square feet of exterior painted surface per dwelling. Also includes soil control work.
Time Limit Per Dwelling	One day	None	None
Resident Location	Inside dwelling but outside work area for duration of project until cleanup has been completed. Alternatively, resident can leave until all work has been completed. Resident must have lead-safe access to entry/egress pathways.	Relocated from dwelling during workday, but may return after daily cleanup has been completed.	Relocated from dwelling for duration of project until final clearance is achieved.

Cont'd

TABLE 2.5 Exterior Worksite Preparation Levels (not including windows)—cont'd

DESCRIPTION	LEVEL 1	LEVEL 2	LEVEL 3
Containment and Barrier System	One layer of plastic on ground extending 10 feet beyond the perimeter of working surfaces. Do not anchor ladder feet on top of plastic (puncture the plastic to anchor ladders securely to ground). For all other exterior plastic surfaces, protect plastic with boards to prevent puncture from falling debris, nails, etc., if necessary. Raise edges of plastic to create a basin to prevent contaminated runoff in the event of unexpected precipitation. Secure plastic to side of building with tape or other anchoring system (no gaps between plastic and building). Weight all plastic sheets down with two-by-fours or similar objects. Keep all windows within 20 feet of working surfaces closed, including windows of adjacent structures.	Same as Level 1	Same as Level 1
Playground Equipment, Toys, Sandbox	Remove all movable items to a 20-foot distance from working surfaces. Items that cannot be readily moved to a 20-foot distance can be sealed with taped plastic sheeting.	Same as Level 1	Same as Level 1

Security	Erect temporary fencing or barrier tape at a 20-foot perimeter around working surfaces (or less if distance to next building or sidewalk is less than 20 feet). If an entryway is within 10 feet of working surfaces, require use of alternative entryway. If practical, install vertical containment to prevent exposure. Use a locked dumpster, covered truck, or locked room to store debris before disposal.	Same as Level 1
Signs	Post warning signs on building and at a 20-foot perimeter around it (or less if distance to next building or sidewalk is >20 feet).	Same as Level 1
Weather	Do not conduct work if wind speeds are <20 miles per hour. Work must stop and cleanup must occur before rain begins.	Same as Level 1
Cleanup	Do not leave debris or plastic out overnight if work is not completed. Keep all debris in secured area until final disposal.	Same as Level 1
Porches	One lead-safe entryway must be made available to residents at all times. Do not treat front and rear porches at the same time if there is not a third doorway.	Front and rear porches can be treated at the same time, unless unprotected workers must use the entryway. Same as Level 2

provide portable showers, which makes worker decontamination more effective and feasible (Figure 2.2). In addition, make sure that all tools are decontaminated after use (Figure 2.3).

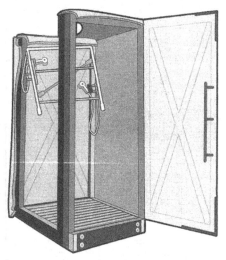

Figure 2.2 Portable shower.

Figure 2.3 Tool decontamination.

What You Need When Working with Lead

3

To a great extent what you need when working with lead depends on what you are doing with lead. The first thing you need is a comprehensive understanding of the laws, rules, and regulations governing the work that you are doing. You will get that in later chapters of this book. Then you will need a basic tool kit. The items are not complex, yet they are needed.

You will also need the EPA pamphlet "Renovate Right: Important Lead Hazard Information for Families, Child Care Providers and Schools." You are required to provide property owners and occupants with this document when you are working with lead in or on their properties. (Current federal regulations require contractors to provide a copy of the *Renovate Right* pamphlet to owners and occupants prior to starting work in pre-1978 housing, as well as to owners and operators of childcare facilities and schools built prior to 1978, and to provide information to parents or guardians of children under age 6 that attend.)

To learn more about the requirements and how to obtain copies of the pamphlet contact the National Lead Information Center at 1-800-424-LEAD (5323) or visit EPA's website at *www.epa.gov/lead*. Some of the other tools you need are listed in Box 3.1. Personal protective equipment needed is shown in Box 3.2.

▶ NEW RULES FOR CONTRACTORS

Beginning April 2010 contractors performing work that disturbs lead-based paint in homes, childcare facilities, and schools built before 1978 must:

- Be EPA certified and follow specific work practices to prevent lead contamination. To learn more about how you can meet these requirements contact the National Lead Information Center at 1-800-424-LEAD (5323) or visit www.epa.gov/lead.

Box 3.1 Tools and Supplies for Working with Lead

- Barriers and signs
- Tape
- Stapler
- Heavy plastic sheeting
- Utility knife or scissors
- Wet/dry sandpaper, sanding sponge
- Misting bottle, pump sprayer
- Chemical stripper
- Power tools with high-efficiency particulate air (HEPA) filter-equipped vacuum attachments
- Low-temperature heat gun
- Heavy-duty plastic bags
- HEPA vacuum cleaner
- Paper towels or disposable wipes
- Mop and disposable mop heads
- General-purpose cleaner
- Buckets
- Shovel and rake

Box 3.2 Lead Personal Protective Equipment

- Eyewear
- Painters' hats
- Gloves
- Coveralls
- Disposable shoe covers
- N-100-rated disposable respirator

- Be prepared for these new requirements. Adopt the following simple practices and you can work safely with lead. Talk to the residents and make them aware of the work you will do and how you will do it. The following are some recommendations:
  - Explain the steps you will take to protect residents from lead.
  - Set up work areas that will not expose residents.
  - Minimize the dust.
  - Leave the work area clean.

Conduct a survey of the homeowner to gain information. See Figure 3.1 for a sample questionnaire that can be used. Consider

Resident Questionnaire
(To be completed by risk assessor via interview with resident.)

Children/Children's Habits

1. (a) Do you have any children that live in your home?   Yes_____   No_____
       (If no children, skip to Question 5.)
   (b) If yes, how many? _____   Ages?   _____ _____ _____ _____ _____
   (c) Record blood-lead levels, if known.   _____ _____ _____ _____ _____
   (d) Are there women of child-bearing age present?   Yes_____   No_____

2. Location of the rooms/areas where each child sleeps, eats, and plays.

Name of child	Location of bedroom	Location of all rooms where child eats	Primary location where child plays *indoors*	Primary location where child plays *outdoors*

3. Where are toys stored/kept? _____

4. Is there any visible evidence of chewed or peeling paint on the woodwork, furniture, or toys?
   Yes_____   No_____

Family Use Patterns

5. Which entrances are used most frequently?   _____
6. Which windows are opened most frequently?   _____
7. Do you use window air conditioners? If yes, where?   _____
   (Condensation often causes paint deterioration)
8. (a) Do any household members garden?   Yes_____   No_____
   (b) Location of garden.   _____
   (c) Are you planning any landscaping activities that will remove
       grass or ground covering?   Yes_____   No_____
9. (a) How often is the household cleaned?   _____
   (b) What cleaning methods do you use?   _____
10. (a) Did you recently complete any building renovations?   Yes_____   No_____
   (b) If yes, where?   _____
   (c) Was building debris stored in the yard? If yes, where?   _____
11. Are you planning any building renovations? If yes, where?   _____
12. (a) Do any household members work in a lead-related industry?   Yes_____   No_____
   (b) If yes, where are dirty work clothes placed and cleaned?   _____

Figure 3.1 Resident questionnaire.

all of the options. Tables 3.1 and 3.2 can help with this. Do your own inspection and look for potential reasons for a problem with lead paint. See Figure 3.2 for locations to check on a home.

Before you begin work, do a risk assessment. Figure 3.3 gives an indication of what this screening entails. After this, consider whether the job is low risk or high risk for surfaces known or suspected to contain lead-based paint. The summary in

TABLE 3.1 Hazard Control Options to Identify during Risk Assessments

TREATMENT OPTION	DUST[1] ON FLOOR	DUST[1] ON WINDOWS	PAINT[2] ON DOORS	PAINT[2] ON WINDOWS	PAINT[2] ON FLOOR AND WALLS	PAINT[2] ON TRIM	HIGH SOIL LEAD LEVELS
Dust removal	X	X	X	X	X	X	X
Paint film stabilization			X	X	X	X	
Friction-reduction treatments	X	X		X		X	
Impact-reduction treatments	X	X	X				
Planting grass	X						X
Planting sod	X						X
Paving the soil	X						X
Encapsulation			X	X	X	X	
Paint removal by heat gun[3]			X	X	X	X	
Paint removal by chemical[3]			X	X	X	X	
Paint removal by contained abrasive[3]						X	
Soil removal	X	X					X
Building component replacement			X	X	X	X	

[1] Lead-contaminated dust

[2] Deteriorated lead-based paint

[3] Limited areas only

TABLE 3.2 Similarities between Lead Hazard Control and Renovation

RENOVATION TECHNIQUE	LEAD HAZARD CONTROL TECHNIQUE
Repainting	Paint film stabilization
Window and door repair	Friction and impact surface treatments
Landscaping	Soil treatment
Installation of new building components (e.g., cabinet replacement)	Building component replacement
Paint stripping	Onsite paint removal
New wall installation	Enclosure

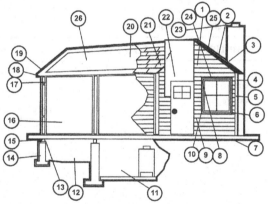

(1) siding exceeds 14% water content; (2) no cricket where chimney meets roof; (3) no step flashing at side of chimney; (4) corner rim not caulked; (5) exposed nailheads rusting; (6) no window wash at window sill; (7) wood contacts earth; (8) no drip or gutter at eaves; (9) poorly fitted window and door trims; (10) water proof paper not installed behind trim; (11) damp, wet cellar unventilated at opposite sides; (12) no ventilation of unexcavated space; (13) no blocking between unexcavated space and stud wall space; (14) no water proofing or drainage tile around cellar walls; (15) no foundation water and termite sill; (16) plaster not dry enough to paint; (17) sheathing paper that is not waterproof; (18) vapor barrier omitted—needed for present or future insulation; (19) roof built during wet, rainy season without taking due precaution or ventilating on dry days; (20) roof leaks; (21) inadequate flashing at breaks, corners, roof; (22) poorly matched joints; (23) no chimney cap; (24) no flashing over openings; (25) full of openings, loosely built; (26) no or inadequate ventilation of attic space

Figure 3.2 Moisture-related causes of paint failure.

Table 3.3 can help with this. Once this is done, you can decide which protective measures to take. Table 3.4 gives an indication of this procedure.

When working in homes, childcare facilities, and schools built before 1978 you must provide the *Renovate Right* pamphlet to

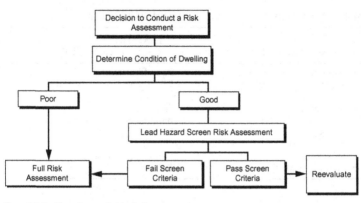

Figure 3.3 Lead hazard screen decision logic.

TABLE 3.3 Low- and High-Risk Job Designations

JOB DESCRIPTION	LOW RISK	HIGH RISK*
Repainting (includes surface preparation)		✓
Plastering or wall repair		✓
Window repair		✓
Window pane or glass replacement only	✓	
Water or moisture damage repair (repainting and plumbing)		✓
Door repair	✓	
Building component replacement		✓
Welding on painted surfaces		✓
Door lock repair or replacement	✓	
Electrical fixture repair	✓	
Floor refinishing		✓
Carpet replacement		✓
Groundskeeping	✓	
Radiator leak repair	✓	
Baluster repair (metal)		✓
Demolition		✓

*High-risk jobs typically disturb more than 2 square feet per room. If these jobs disturb less than 2 square feet, then they can be considered low-risk jobs.

TABLE 3.4 Protective Measures for Low- and High-Risk Jobs

PROTECTIVE MEASURE	LOW RISK	HIGH RISK
Worksite preparation with plastic sheeting (6-mil thick)	Plastic sheet no less than 5 feet by 5 feet immediately underneath work area	Whole floor, plus simple airlock at door or tape door shut
Children kept out of work area	Yes	Yes
Resident relocation during work	No	Yes
Respirators	Probably not necessary*	Recommended
Protective clothing *Note:* Protective shoe coverings are not to be worn on ladders, scaffolds, etc.	Probably not necessary*	Recommended
Personal hygiene (enforced hand washing after job)	Required	Required
Showers	Probably not necessary	Recommended
Work practices	Use wet methods, except near electrical circuits	Use wet methods, except near electrical circuits
Cleaning	Wet cleaning with lead-specific detergent, trisodium phosphate, or other suitable detergent around the work area only (2 linear feet beyond plastic)	HEPA vacuum/wet wash/HEPA vacuum the entire work area
Clearance	Visual examination only	Dust sampling during the preliminary phase of the maintenance program and periodically thereafter (not required for every job)

*Employers must have objective data showing that worker exposures are less than the OSHA permissible exposure limit of 50 μg/m³ if respirators and protective clothing will not be provided.

residents or the facility operator before the job begins. You must also provide information to families whose children attend the childcare facility or school.

▶ SAFE WORK AREAS

Creating safe work areas is an important part of working with lead. There are many things you can do to increase safety and reduce potential problems. Start by selecting appropriate personal protective equipment. Then make sure that your workers use it. This should include appropriate eyewear, clothing, and respiratory protection for the job.

The work area should be contained so that no dust or debris leaves the work area. Use signs to keep residents and pets out of the work area. Remove furniture and belongings, or cover them securely with heavy plastic sheeting. Use heavy plastic sheeting to cover floors and other fixed surfaces such as large appliances in the work area. When appropriate, use heavy plastic sheeting to separate the work area from the rest of the residence. Close and seal vents in the work area and, if necessary, turn off forced-air heating and air-conditioning systems.

What should you do outside to create a safe work site? Mark off the work area to keep nonworkers away. Cover the ground and plants with heavy plastic sheeting. Close windows and doors near the work area. Move or cover play areas located near the work area.

Minimize the Dust

Dust containing lead is dangerous. You need to minimize it. How can you do this? Mist areas before sanding, scraping, drilling, and cutting. The water will control dust to a great extent. Score paint before separating components. Pry and pull apart components instead of pounding and hammering. The beating of materials containing lead is more likely to generate dust. Always use a shroud with HEPA vacuum attachment when using power tools and equipment. See the following tables and figures for help in dealing with dust. These include Tables 3.5 and 3.6 along with Box 3.3 and Figure 3.4.

Dangerous Practices

There are some dangerous practices when working with lead that you should simply avoid. For example, do not use an open flame, burning, or torching to remove material containing

TABLE 3.5 Potential Sources of Lead-Contaminated House Dust

SOURCE	WHAT CONTRIBUTES TO LEAD IN DUST	KEY SITES
Interior lead-based paint	Deteriorating paint Friction/abrasion Impact Water damage Planned disturbances: maintenance activities, repainting, remodeling, abatement	All surfaces Windows, doors, stairs, and floors Door systems, openings, baseboards, corner edges, chair rails, and stair risers Walls, trim, and ceilings All surfaces coated with lead-based paint
Exterior lead-based paint	Tracking (by humans and pets) and blowing of leaded dust from weathered, chalked, or deteriorated exterior lead-based paint; also direct contact with such paint Demolition and other disturbances of lead-based paint on buildings and nearby steel structures	All exterior lead-based painted components, including porches and window sills Exposed soil, sandboxes, side walks, and window troughs
Soil and exterior dust	Tracking (by humans and pets) and blowing of exterior soil/dirt contaminated with lead from deteriorating exterior lead-based paint; past deposition of lead in gasoline	Exposed soil, sandboxes, side walks, streets, and window troughs
Point sources	Releases from lead-related industries (i.e., smelters, battery recycling, incinerators)	Location of point sources
Hobby activities	Cutting, molding, and melting of lead for bullets, fishing sinkers, toys, and joining stained glass. Use of lead-containing glazes and paints. Restoration of lead-based painted items	Rooms in which hobbies are pursued
Occupational sources	Transport of lead-contaminated dust from the job to home on clothing, tools, hair, and vehicles	Vehicles and laundry rooms, changing areas, furniture, and entryway rugs

TABLE 3.6 Major Dust Reservoirs and Potential Dust Traps

INTERIOR	EXTERIOR
Window sills	Porch systems
Floors/steps	Window troughs
Cracks and crevices	Steps
Carpets and rugs	Exposed soil
Mats	Sandboxes
Upholstered furnishings	
Window coverings	
Radiators	
Grates and registers	
Heating, ventilation, air conditioning filters	

Box 3.3 Example of Dust Sampling Plan

Dust samples should be collected from each of the following locations:

- One from the floor of the child's principal play area, TV room, or living room.
- One from the interior window sill of the most frequently opened window in the child's principal play area.
- One from the floor of the kitchen.
- One from kitchen window trough (if inaccessible, an interior window sill sample should be collected).
- One from bedroom floor of the youngest child (older than 6 months).
- One from the interior window sill of the bedroom of the youngest child (older than 6 months).
- One from bedroom floor of the next oldest child, if any.
- One from the window trough of the bedroom of the next oldest child, if any (if inaccessible, an interior window sill sample should be collected).

At least one window trough sample should be collected in each dwelling. If no playroom can be identified, the living room should be sampled. If the youngest child's bedroom cannot be identified, the smallest bedroom should be sampled. .

Under this plan, three composite samples *or* eight single-surface samples would be collected. The risk assessor should use professional judgment to determine which method is most appropriate.

In some dwellings, it may be appropriate to delete or add a sample location. For example, if a window is never opened, the window trough should not be sampled. If an additional location is identified that displays both a visible accumulation of dust *and* has obviously been exposed to a child, an additional sample from that location should be collected. A dusty tabletop in the child's play area, or a cabinet with deteriorated paint that holds dishes, are surfaces that should be sampled.

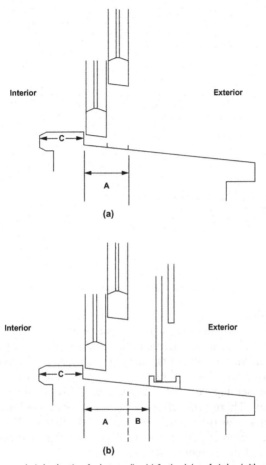

Figure 3.4 Suggested window locations for dust sampling. (a). Sectional view of window (with no strom window) showing window trough area, A, to be tested. Trough is the surface where both window sashes can touch the sill when lowered. The interior window sill (stool) is shown as area C. Interior window sills and window troughs should be sampled separately. (b). Sectional view of window (including storm window) showing window trough area, A and B, to be tested. Trough extends out to storm window frame. The interior window sill (stool) is shown as area C. Interior window sills and window troughs should be sampled separately.

lead. Sanding, grinding, planing, needle gunning, and blasting with power tools is not recommended unless equipped with a shroud and HEPA vacuum attachment. Don't use a heat gun at temperatures greater than 1100°F for the removal of lead-containing materials.

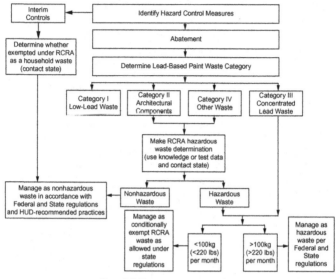

Figure 3.5 Waste management procedures.

On a Daily Basis

On a daily basis there are certain tasks that you should ensure are done. Put trash and debris in heavy-duty plastic bags. Wrap waste building components, such as windows and doors, in heavy plastic sheeting and tape shut. You need a waste management plan; see Figure 3.5 for an example of waste management procedures. There are different categories of hazardous waste. Once you know the categories you are dealing with, you can determine a plan of action. See Table 3.7 for clarification on the management of abatement waste.

Ensure that everything, including tools, equipment, and even workers, are free of dust and debris before leaving the work area. HEPA vacuum the work area. Wash up and change out of work clothes before you and your workers go home. Remember, you do not want to bring lead-based paint dust home and expose your family. Remind residents to stay out of the work area.

▶ WHEN A JOB IS COMPLETE

When a job is complete there is a routine list of tasks to be done. For example, remove the plastic sheeting carefully, mist with water, fold dirty side in, tape shut, and dispose of it. HEPA

TABLE 3.7 Categories of Abatement Waste

CATEGORY	DESCRIPTION	EXAMPLES OF WASTES
I	Low-Lead Waste	• Filtered personal and commercial wash water • Disposable personal protective clothing that has been HEPA vacuumed before disposal • Plastic sheeting cleaned prior to disposal (misted and wiped) and carpeting • Any waste that is determined to be nonhazardous by TCLP testing and is not an EPA-listed hazardous waste
II	Architectural Components	Painted finish carpentry items; for example: • Doors • Windows • Window trim and sills • Baseboards • Railings • Moldings Other painted building components; for example: • Metal railings • Radiators • Walls • Stone or brick
III	Concentrated Lead Waste	Sludge from paint stripping Lead-based paint chips and dust HEPA vacuum debris and filter Unfiltered wash water Hazardous waste Any waste included on EPA's list of hazardous waste
IV	Other Waste	Material that cannot be determined, using knowledge of the waste, to be either hazardous or nonhazardous must be tested using TCLP

vacuum all surfaces, including walls. Wash the work area with a general-purpose cleaner. Check your work carefully for lead dust because hazardous amounts may be minute and not easily visible. If you see any dust or debris, then clean the area again.

Perform a final clean-up check. Use disposable cleaning cloths to wipe the floor of the work area and compare them to a cleaning verification card to determine if the work area was adequately cleaned. A cleaning verification card with detailed instructions can be ordered from the EPA website, *www.epa.gov/lead*, or from the National Lead Information Center, 1-800-424-LEAD (5323).

You now have a good idea of what you need to work with lead, including some suggestions for recommended work practices. Now, let's move on.

OSHA Requirements for Lead Construction

4

Any experienced contractor is aware of the purpose of the Occupational Safety and Health Administration (OSHA). Part of the U.S. Department of Labor, it is the government agency that makes rules regarding safety in the workplace and levies stiff fines for those who abuse those rules. No contractor wants to be on the wrong side of OSHA. Failing a code enforcement inspection is bad. Failing an OSHA inspection is worse, and more expensive.

When it comes to working with hazardous materials, both OSHA and the EPA are actively involved in regulations. This chapter focuses on the OSHA requirements. It is unlikely that you will memorize them all, but you should become familiar with the rules. The reading here is not particularly entertaining, but it is enlightening. At the very least, you need to learn your way around the facts so that you can stay out of trouble on a job site. The information that follows is based on OSHA's view of working with lead on construction sites.

▶ DEFINITIONS

The following are OSHA definitions for terms used when working with lead-based paint and related materials.

- *Action level* means employee exposure, without regard to the use of respirators, to an airborne concentration of lead of 30 micrograms per cubic meter of air ($30\,\mu g/m^3$) calculated as an 8-hour time-weighted average (TWA).
- *Assistant Secretary* means the Assistant Secretary of Labor for Occupational Safety and Health, U.S. Department of Labor, or designee.
- *Competent person* means one who is capable of identifying existing and predictable lead hazards in the surroundings or working conditions and who has authorization to take prompt corrective measures to eliminate them.

- *Director* means the Director, National Institute for Occupational Safety and Health (NIOSH), U.S. Department of Health and Human Services, or designee.
- *Lead* means metallic lead, all inorganic lead compounds, and organic lead soaps. Excluded from this definition are all other organic lead compounds.

► PERMISSIBLE EXPOSURE LIMIT

The employer shall ensure that no employee is exposed to lead at concentrations greater than 50 micrograms per cubic meter of air ($50\,\mu g/m^3$) averaged over an 8-hour period.

If an employee is exposed to lead for more than 8 hours in any work day the employee's allowable exposure, as a TWA for that day, shall be reduced according to the following formula:

$$\text{Allowable employee exposure (in}\,\mu g\,/\,m^3) = 400 \div \text{hours worked in the day}$$

See Table 4.1 for exposure limits for both lead in the air and in blood. Table 4.2 shows the OSHA standard by exposure level.

When respirators are used to limit employee exposure as required under OSHA Standard 1926.62 paragraph (c) of this section and all the requirements of paragraphs (e)(1) and (f) of this section have been met, employee exposure may be considered to be at the level provided by the protection factor of the respirator for those periods the respirator is worn. Figure 4.1 shows a worker in protective clothing and wearing a

TABLE 4.1 Worker Exposure Limits and Guidelines

AIR LEAD	
OSHA Permissible Exposure Limit (PEL)	$50\,\mu g/m^3$
OSHA action level	$30\,\mu g/m^3$
BLOOD LEAD	
OSHA medical removal limit	$50\,\mu g/dL$
OSHA recommended level to prevent reproductive problems	$30\,\mu g/dL$
ACGIH proposed threshold limit value	$20\,\mu g/dL$

TABLE 4.2 Required Action under the OSHA Standard by Exposure Level

CATEGORY I	CATEGORY II	CATEGORY III
30 µg/m³* and under (below the action level)	30–50 µg/m³ (above the action level, but below the PEL)	50 µg/m³ and over (above the PEL)
Train employees Conduct exposure monitoring Maintain records	Same as category I, plus: <ul><li>Provide respirator at employee request</li><li>Conduct exposure monitoring every 3 months</li><li>Conduct blood lead monitoring</li></ul>	Same as category II, plus: <ul><li>Enforce respirator use</li><li>Enforce use of protective clothing</li><li>Develop monitoring every 6 months</li><li>Enforce housekeeping</li><li>Provide hygiene facilities and enforce washing</li></ul>

*All exposure levels are 8-hour, time-weighted averages.

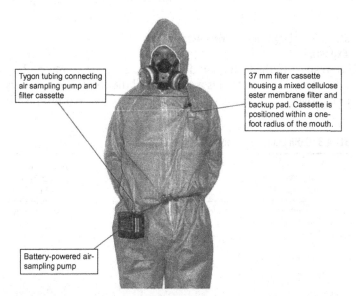

Figure 4.1 Worker dressed in protective clothing and wearing a respirator.
Note: Personal breaking zone air sampling involves drawing air through a 0.8u mixed-cellulose ester filter in a close-face 37 mm filter cassette using an air-sampling pump running at 2 liters/minute.

respirator. Those periods may be averaged with exposure levels during periods when respirators are not worn to determine the employee's daily TWA exposure.

Exposure Assessment

Each employer who has a workplace or operation covered by this standard shall initially determine if any employee may be exposed to lead at or above the action level. For the purposes of paragraph (d) of this section, employee exposure is the exposure that would occur if the employee were not using a respirator.

With the exception of monitoring under paragraph (d)(3), where monitoring is required under this section, the employer shall collect personal samples representative of a full shift including at least one sample for each job classification in each work area either for each shift or for the shift with the highest exposure level. See Table 4.3 for an example of OSHA's lead hazard control assumed exposures. Full shift personal samples shall be representative of the monitored employee's regular, daily exposure to lead.

Protection of Employees during Assessment of Exposure

With respect to the lead related tasks listed in paragraph (d)(2)(i) of this section, where lead is present, until the employer performs an employee exposure assessment as required in paragraph (d)

TABLE 4.3   OSHA Lead Hazard Control Assumed Exposures

50 µG/M³ TO 500 µG/M³	500 µG/M³ TO 2,500 µG/M³	GREATER THAN 2,500 µG/M³
Manual demolition	Cleanup on dry, abrasive blasting jobs	Abrasive blasting
Manual scraping	Abrasive blasting enclosure movement/ removal	
Manual sanding		
Heat gun use		
Power tool paint removal in the HEPA vacuum-assist dust collection system		

Note: Abrasive blasting without a HEPA local exhaust system is not permitted in residential dwellings.
Adapted from 29 CFR 1926.62.

of this section and documents that the employee performing any of the listed tasks is not exposed above the PEL, the employer shall treat the employee as if the employee were exposed above the PEL, and not in excess of ten (10) times the PEL, and shall implement employee protective measures prescribed in paragraph (d)(2)(v) of this section.

The tasks covered by this requirement are:

- Where lead-containing coatings or paint are present: manual demolition of structures (e.g., dry wall), manual scraping, manual sanding, heat gun applications, and power tool cleaning with dust collection systems; spray painting with lead paint
- In addition, with regard to tasks not listed in paragraph (d)(2)(i), where the employer has any reason to believe that an employee performing the task may be exposed to lead in excess of the PEL, until the employer performs an employee exposure assessment as required by paragraph (d) of this section and documents that the employee's lead exposure is not above the PEL, the employer shall treat the employee as if the employee were exposed above the PEL and shall implement employee protective measures as prescribed in paragraph (d)(2)(v) of this section.
- With respect to the tasks listed in paragraph (d)(2)(iii) of this section, where lead is present, until the employer performs an employee exposure assessment as required in paragraph (d) of this section, and documents that the employee performing any of the listed tasks is not exposed in excess of $500\,\mu g/m^3$, the employer shall treat the employee as if the employee were exposed to lead in excess of $500\,\mu g/m^3$ and shall implement employee protective measures as prescribed in paragraph (d)(2)(v) of this section. Where the employer does establish that the employee is exposed to levels of lead below $500\,\mu g/m^3$, the employer may provide the exposed employee with the appropriate respirator prescribed for such use at such lower exposures. The tasks covered by this requirement are using lead containing mortar or lead burning
- Where lead-containing coatings or paint are present: rivet busting; power tool cleaning without dust collection systems; cleanup activities where dry expendable abrasives are used; and abrasive blasting enclosure movement and removal

- With respect to the tasks listed in this paragraph (d)(2)(iv) of this section, where lead is present, until the employer performs an employee exposure assessment as required in paragraph (d) of this section and documents that the employee performing any of the listed tasks is not exposed to lead in excess of $2500 \mu g/m^3$ (50 × PEL), the employer shall treat the employee as if the employee were exposed to lead in excess of $2500 \mu g/m^3$ and shall implement employee protective measures as prescribed in paragraph (d)(2)(v) of this section. Where the employer does establish that the employee is exposed to levels of lead below $2500 \mu g/m^3$, the employer may provide the exposed employee with the appropriate respirator prescribed for use at such lower exposures. Interim protection as described in this paragraph is required where lead-containing coatings or paint are present on structures when performing abrasive blasting, welding, cutting, and torch burning.
- Until the employer performs an employee exposure assessment as required under paragraph (d) of this section and determines actual employee exposure, the employer shall provide to employees performing the tasks described in paragraphs (d)(2)(i), (d)(2)(ii), (d)(2)(iii), and (d)(2)(iv) of this section interim protection as follows:

Appropriate Respiratory Protection

Except as provided under paragraphs (d)(3)(iii) and (d)(3)(iv) of this section, the employer shall monitor employee exposures and shall base initial determinations on the employee exposure monitoring results and any of the following, relevant considerations:

- Any information, observations, or calculations that would indicate employee exposure to lead
- Any previous measurements of airborne lead
- Any employee complaints of symptoms that may be attributable to exposure to lead

Monitoring for the initial determination where performed may be limited to a representative sample of the exposed employees who the employer reasonably believes are exposed to the greatest airborne concentrations of lead in the workplace.

Where the employer has previously monitored for lead exposures, and the data were obtained within the past 12 months during work operations conducted under workplace conditions closely resembling the processes, type of material, control methods, work practices, and environmental conditions used and prevailing in the employer's current operations, the employer may rely on such earlier monitoring results to satisfy the requirements of paragraphs (d)(3)(i) and (d)(6) of this section if the sampling and analytical methods meet the accuracy and confidence levels of paragraph (d)(10) of this section.

Where the employer has objective data, demonstrating that a particular product or material containing lead or a specific process, operation, or activity involving lead cannot result in employee exposure to lead at or above the action level during processing, use, or handling, the employer may rely on such data instead of implementing initial monitoring.

The employer shall establish and maintain an accurate record documenting the nature and relevancy of objective data as specified in paragraph (n)(4) of this section, where used in assessing employee exposure in lieu of exposure monitoring. Objective data, as described in paragraph (d)(3)(iv) of this section, is not permitted to be used for exposure assessment in connection with paragraph (d)(2) of this section.

You need an evaluation plan before you send workers into a job. Figure 4.2 shows suggestions for such a program. If you are dealing with a targeted risk building, the size of the structure is a factor in your risk assessment. Study Table 4.4 for advice on this.

Positive Initial Determination and Initial Monitoring

Where a determination conducted under paragraphs (d)(1), (2), and (3) of this section shows the possibility of any employee exposure at or above the action level, the employer shall conduct monitoring that is representative of the exposure for each employee in the workplace who is exposed to lead.

Where the employer has previously monitored for lead exposure, and the data were obtained within the past 12 months during work operations conducted under workplace conditions closely resembling the processes, type of material, control methods, work practices, and environmental conditions used and prevailing in the employer's current operations, the employer may rely on such earlier monitoring results to satisfy the requirements of paragraph (d)(4)(i) of this section if the sampling and analytical methods meet the accuracy and confidence levels of paragraph (d)(10) of this section.

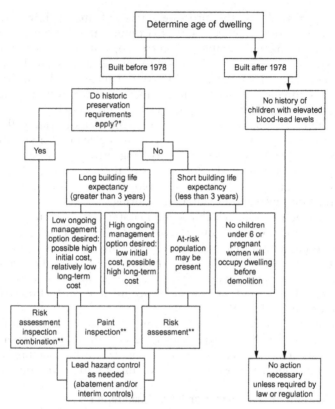

Figure 4.2 Risk assessment versus inspection—a decision-making grid.

*Whenever it is determined that a child with an elevated blood-lead level is living in the unit, a lead-hazard investigation must be completed.

**This is only general guidance. Actual on-site conditions or regulatory requirements may dictate another hazard evaluation method. A paint inspection by itself may not identify lead-based paint hazards. A risk assessment inspection combination is an option whenever an assessment or inspection is indicated. A risk assessment screen is appropriate for buildings in good condition. Some jurisdictions may limit choices in some circumstances.

Negative Initial Determination

Where a determination, conducted under paragraphs (d)(1), (2), and (3) of this section, is made that no employee is exposed to airborne concentrations of lead at or above the action level, the employer shall make a written record of such determination. The record shall include at least the information specified in paragraph (d)(3)(i) of this section and shall also include the date

TABLE 4.4 Risk Assessment Approach for Different Size Evaluations

ACTION REQUIRED	OWNER-OCCUPIED, SINGLE-FAMILY DWELLINGS	FIVE OR MORE SIMILAR RENTAL DWELLINGS	LESS THAN FIVE RENTAL DWELLINGS OR RENTAL DWELLINGS THAT ARE NOT SIMILAR
Assess every dwelling	Yes	No	Yes*
Deteriorated paint sampling (if no inspection conducted)	Yes	Yes	Yes
Dust sampling	Yes	Yes	Yes
Bare soil sampling	Yes	Yes	Yes
Water sampling	Optional	Optional	Optional
Air sampling	No	No	No
Management system analysis	Not applicable	Optional	Optional
Maintenance work systems modified	Cleaning and repair practices modified	Optional	Optional
Housing condition and characteristics assessment	Yes	Yes	Yes
Demographics and use patterns description	Yes	Yes	Yes

*There may be occasions when it is not necessary to sample all nonsimilar dwellings.

of determination, location within the worksite, and the name and social security number of each employee monitored.

Frequency
If the initial determination reveals employee exposure to be below the action level, further exposure determination need not be repeated except as otherwise provided in paragraph (d)(7) of this section.

If the initial determination or a subsequent determination reveals employee exposure to be at or above the action level but at or below the PEL, the employer shall perform monitoring in accordance with this paragraph at least every 6 months. The employer shall continue monitoring at the required frequency until at least two consecutive measurements, taken at least 7 days apart, are below the action level at which time the employer may discontinue monitoring for that employee except as otherwise provided in paragraph (d)(7) of this section.

If the initial determination reveals that employee exposure is above the PEL, the employer shall perform monitoring quarterly. The employer shall continue monitoring at the required frequency until at least two consecutive measurements, taken at least 7 days apart, are at or below the PEL but at or above the action level, at which time the employer shall repeat monitoring for that employee at the frequency specified in paragraph (d)(6)(ii) of this section, except as otherwise provided in paragraph (d) (7) of this section. The employer shall continue monitoring at the required frequency until at least two consecutive measurements, taken at least 7 days apart, are below the action level, at which time the employer may discontinue monitoring for that employee except as otherwise provided in paragraph (d)(7) of this section.

Additional Exposure Assessments

Whenever there has been a change of equipment, process, control, or personnel or a new task has been initiated that may result in additional employees being exposed to lead at or above the action level or may result in employees already exposed at or above the action level being exposed above the PEL, the employer shall conduct additional monitoring in accordance with this paragraph.

Employee Notification

The employer must, as soon as possible but no later than 5 working days after the receipt of the results of any monitoring performed under this section, notify each affected employee of these results either individually in writing or by posting the results in an appropriate location that is accessible to employees.

Whenever the results indicate that the representative employee exposure, without regard to respirators, is at or above the PEL, the employer shall include in the written notice a statement that the employee's exposure was at or above that level and a description of the corrective action taken or to be taken to reduce exposure to below that level.

Accuracy of Measurement

The employer shall use a method of monitoring and analysis that has an accuracy (to a confidence level of 95%) of not less than plus or minus 25% for airborne concentrations of lead equal to or greater than $30\,\mu g/m^3$.

▶ METHODS OF COMPLIANCE

The employer shall implement engineering and work practice controls, including administrative controls, to reduce and maintain employee exposure to lead to or below the permissible exposure limit to the extent that such controls are feasible. Wherever all feasible engineering and work practices controls that can be instituted are not sufficient to reduce employee exposure to or below the permissible exposure limit prescribed in paragraph (c) of this section, the employer shall nonetheless use them to reduce employee exposure to the lowest feasible level and shall supplement them by the use of respiratory protection that complies with the requirements of paragraph (f) of this section.

Compliance Program

Prior to commencement of the job, each employer shall establish and implement a written compliance program to achieve compliance with paragraph (c) of this section. Written plans for these compliance programs shall include at least the following:

- A description of each activity in which lead is emitted; e.g., equipment used, material involved, controls in place, crew size, employee job responsibilities, operating procedures and maintenance practices
- A description of the specific means that will be employed to achieve compliance and, where engineering controls are required, engineering plans and studies used to determine methods selected for controlling exposure to lead
- A report of the technology considered in meeting the PEL
- Air-monitoring data that documents source of lead emissions

- A detailed schedule for program implementation, including documentation such as copies of purchase orders for equipment, construction contracts, etc.
- A work practice program that includes items required under paragraphs (g), (h), and (i) of this section and incorporates other relevant work practices such as those specified in paragraph (e)(5) of this section
- An administrative control schedule required by paragraph (e)(4) of this section, if applicable
- A description of arrangements made among contractors on multi-contractor sites with respect to informing affected employees of potential exposure to lead and with respect to responsibility for compliance with this section as set forth in Standard 1926.16
- The compliance program shall provide for frequent and regular inspections of job sites, materials, and equipment to be made by a competent person
- Written programs shall be submitted on request to any affected employee or authorized employee representatives, to the Assistant Secretary and the Director, and shall be available at the worksite for examination and copying by the Assistant Secretary and the Director
- Written programs must be revised and updated at least annually to reflect the current status of the program

Mechanical Ventilation
When ventilation is used to control lead exposure, the employer shall evaluate the mechanical performance of the system in controlling exposure as necessary to maintain its effectiveness.

Administrative Controls
If administrative controls are used as a means of reducing employees' TWA exposure to lead, the employer shall establish and implement a job rotation schedule that includes:

- Name or identification number of each affected employee
- Duration and exposure levels at each job or work station where each affected employee is located
- Any other information that may be useful in assessing the reliability of administrative controls to reduce exposure to lead

- The employer shall ensure that, to the extent relevant, employees follow good work practices

▶ RESPIRATORY PROTECTION

For employees who use respirators required by this section, the employer must provide each employee an appropriate respirator that complies with the following requirements:

- Respirators must be used during periods when an employee's exposure to lead exceeds the PEL
- Work operations for which engineering and work-practice controls are not sufficient to reduce employee exposures to or below the PEL
- Periods when an employee requests a respirator
- Periods when respirators are required to provide interim protection of employees while they perform the operations specified in paragraph (d)(2) of this section

Respirator Program

The employer must implement a respiratory protection program in accordance with §1910.134(b) through (d) (except (d)(1)(iii)), and (f) through (m), which covers each employee required by this section to use a respirator.

If an employee has breathing difficulty during fit testing or respirator use, the employer must provide the employee with a medical examination in accordance with paragraph (j)(3)(i)(B) of this section to determine whether or not the employee can use a respirator while performing the required duty.

Respirator Selection

Employers must select, and provide to employees, the appropriate respirators specified in paragraph (d)(3) of 29 CFR 1910.134. They must provide employees with a full-facepiece respirator instead of a half-mask respirator for protection against lead aerosols that may cause eye or skin irritation at the use concentrations. HEPA filters for powered and nonpowered air-purifying respirators must also be provided.

See Figure 4.3 for different types of respirators used in residential lead hazard control work. Figure 4.4 shows a worker in protective clothing while cleaning. It is important to vacuum

Half-Mask, Air Purifying Respirator

Adequate for atmospheres up to 500 µg/m³ lead

Full-Face, Air Purifying Respirator

Adequate for atmospheres up to 2500 µg/m³ lead

Powered-Air Purifying Respirator

Adequate for atmospheres up to 2500 µg/m³ lead

(filter and battery-powered blower are worn on belt)

Figure 4.3 Types of respirators.

all surfaces in the work area, including those that have been covered with plastic. The employer must provide a powered air-purifying respirator when an employee chooses to use such a respirator and it will provide adequate protection to the employee.

▶ PROTECTIVE WORK CLOTHING AND EQUIPMENT

Where an employee is exposed to lead above the PEL without regard to the use of respirators, where employees are exposed to lead compounds which may cause skin or eye irritation (e.g., lead arsenate, lead azide), and as interim protection for employees performing tasks as specified in paragraph (d)(2)

Figure 4.4 Worker wearing protective clothing during cleaning.

Note: Start at the far end and work toward the decontamination area. Begin with ceilings or the top of walls and work down, cleaning the floor last. Do every inch of windows, especially troughs. Use a corner tool to clean where floor meets baseboard and all the cracks in floor boards. Use the brush tool for walls and move slowly and carefully to get all the dust.

of this section, the employer shall provide at no cost to the employee and ensure that the employee uses appropriate protective work clothing and equipment that prevents contamination of the employee and the employee's garments such as, but not limited to:

- Coveralls or similar full-body work clothing
- Gloves, hats, and shoes or disposable shoe coverlets
- Face shields, vented goggles, or other appropriate protective equipment that complies with OSHA Standard 1910.133

The employer shall provide the protective clothing required in paragraph (g)(1) of this section in a clean and dry condition at least weekly, and daily to employees whose exposure levels without regard to a respirator are over 200 µg/m³ of lead as an 8-hour TWA. See Figure 4.5 for an example of some of the gear that should be available to workers who are dealing with lead.

Cleaning and Replacement

The employer shall provide for the cleaning, laundering, and disposal of protective clothing and equipment required by paragraph (g)(1) of this section. The employer shall repair or replace required protective clothing and equipment as needed to maintain their effectiveness. The employer shall ensure that all protective clothing is removed at the completion of a work shift only in change areas provided for that purpose as prescribed in paragraph (i)(2) of this section.

The employer shall ensure that contaminated protective clothing which is to be cleaned, laundered, or disposed of, is placed in a closed container in the change area, which prevents dispersion of lead outside the container.

The employer shall inform in writing any person who cleans or launders protective clothing or equipment of the potentially harmful effects of exposure to lead. The employer shall ensure

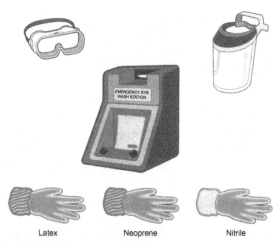

Figure 4.5 Goggles, face shields, gloves, and eye wash facilities.

that the containers of contaminated protective clothing and equipment required by paragraph (g)(2)(v) of this section are labeled as follows:

> **Caution:** *Clothing contaminated with lead. Do not remove dust by blowing or shaking. Dispose of lead contaminated wash water in accordance with applicable local, state, or federal regulations.*

The employer shall prohibit the removal of lead from protective clothing or equipment by blowing, shaking, or any other means that disperses lead into the air.

Housekeeping

All surfaces shall be maintained as free as practicable of accumulations of lead. Clean-up of floors and other surfaces where lead accumulates shall, wherever possible, be cleaned by vacuuming or other methods that minimize the likelihood of lead becoming airborne.

Shoveling, dry or wet sweeping, and brushing may be used only where vacuuming or other equally effective methods have been tried and found not to be effective. Where vacuuming methods are selected, the vacuums shall be equipped with HEPA filters and used and emptied in a manner that minimizes the reentry of lead into the workplace.

Compressed air shall not be used to remove lead from any surface unless the compressed air is used in conjunction with a ventilation system designed to capture the airborne dust created by the compressed air. Figure 4.6 shows the parts of a vacuum with a HEPA filter that is intended for use in lead-hazard conditions. Figure 4.7 shows an example of proper cleaning procedures.

► HYGIENE FACILITIES AND PRACTICES

The employer shall ensure that in areas where employees are exposed to lead above the PEL without regard to the use of respirators, food or beverage is not present or consumed, tobacco products are not present or used, and cosmetics are not applied.

The employer shall provide clean change areas for employees whose airborne exposure to lead is above the PEL, and as interim protection for employees performing tasks as specified in paragraph (d)(2) of this section, without regard to the use of respirators.

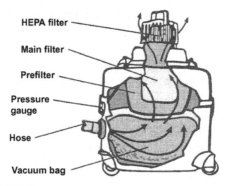

HEPA filter

Main filter

Prefilter

Pressure gauge

Hose

Vacuum bag

Figure 4.6 Parts of a HEPA vacuum.

Note: Most HEPA vacuums suck debris through the hose into the bag, then the air and dust get filtered through the prefilter, the main filter, and the HEPA filter. The lead dust is captured before the air is released into the work area again.

Wet HEPA vacuum

Cleaning agent container

Cleaning agent dispenser and vacuum nozzle

Figure 4.7 Single-pass HEPA vacuum and wet wash technology.

The employer shall ensure that change areas are equipped with separate storage facilities for protective work clothing and equipment and for street clothes, to prevent cross-contamination. The employer shall ensure that employees do not leave the workplace wearing any protective clothing or equipment that is required to be worn during the work shift.

The employer shall provide shower facilities, where feasible, for use by employees whose airborne exposure to lead is above the PEL. The employer shall ensure, where shower facilities are available, that employees shower at the end of the work shift and shall provide an adequate supply of cleansing agents and towels for use by affected employees.

Eating Facilities and Practices

The employer shall provide lunchroom facilities or eating areas for employees whose airborne exposure to lead is above the PEL, without regard to the use of respirators. The employer shall ensure that lunchroom facilities or eating areas are as free as practicable from lead contamination and are readily accessible to employees.

The employer shall ensure that employees whose airborne exposure to lead is above the PEL, without regard to the use of a respirator, wash their hands and face prior to eating, drinking, smoking, or applying cosmetics.

The employer shall ensure that employees do not enter lunchroom facilities or eating areas with protective work clothing or equipment unless surface lead dust has been removed by vacuuming, downdraft booth, or other cleaning method that limits dispersion of lead dust. The employer shall provide adequate handwashing facilities for use by employees exposed to lead in accordance with 29 CFR 1926.51(f).

Where showers are not provided, the employer shall ensure that employees wash their hands and face at the end of the work-shift.

▶ MEDICAL SURVEILLANCE

The employer shall make available initial medical surveillance to employees occupationally exposed on any day to lead at or above the action level. Initial medical surveillance consists of biological monitoring in the form of blood sampling and analysis for lead and zinc protoporphyrin levels.

The employer shall institute a medical surveillance program in accordance with paragraphs (j)(2) and (j)(3) of this section for all employees who are or may be exposed by the employer at or above the action level for more than 30 days in any consecutive 12 months.

The employer shall ensure that all medical examinations and procedures are performed by or under the supervision of a licensed physician. The employer shall make available the required medical surveillance including multiple physician review under paragraph (j)(3)(iii) without cost to employees and at a reasonable time and place.

Biological Monitoring

The employer shall make available biological monitoring in the form of blood sampling and analysis for lead and zinc protoporphyrin levels to each employee covered under paragraphs (j)(1)(i) and (ii) of this section on the following schedule: For each employee covered under paragraph (j)(1)(ii) of this section, at least every 2 months for the first 6 months and every 6 months thereafter.

For each employee covered under paragraphs (j)(1)(i) or (ii) of this section whose last blood sampling and analysis indicated a blood-lead level at or above 40 µg/dL, at least every 2 months. This frequency shall continue until two consecutive blood samples and analyses indicate a blood-lead level below 40 µg/dL and for each employee who is removed from exposure to lead due to an elevated blood-lead level at least monthly during the removal period.

Whenever the results of a blood-lead level test indicate that an employee's blood-lead level exceeds the numerical criterion for medical removal under paragraph (k)(1)(i) of this section, the employer shall provide a second (follow-up) blood sampling test within 2 weeks after the employer receives the results of the first blood sampling test.

Blood-lead level sampling and analysis provided pursuant to this section shall have an accuracy (to a confidence level of 95%) within plus or minus 15% or 6 µg/dL, whichever is greater, and shall be conducted by a laboratory approved by OSHA.

Employee Notification

Within 5 working days after the receipt of biological monitoring results, the employer shall notify each employee in writing of his or her blood-lead level and the employer shall notify each employee whose blood-lead level exceeds 40 µg/dL that the standard requires temporary medical removal with Medical Removal Protection benefits when an employee's blood-lead level exceeds the numerical criterion for medical removal under paragraph (k)(1)(i) of this section.

The employer shall make available medical examinations and consultations to each employee covered under paragraph (j)(1)(ii) of this section on the following schedule:

- At least annually for each employee for whom a blood sampling test conducted at any time during the preceding 12 months indicated a blood-lead level at or above 40 μg/dL as soon as possible, upon notification by an employee either that the employee has developed signs or symptoms commonly associated with lead intoxication, that the employee desires medical advice concerning the effects of current or past exposure to lead on the employee's ability to procreate a healthy child, that the employee is pregnant, or that the employee has demonstrated difficulty in breathing during a respirator fitting test or during use.
- As medically appropriate for each employee either removed from exposure to lead due to a risk of sustaining material impairment to health, or otherwise limited pursuant to a final medical determination.

▶ MEDICAL MATTERS

The content of medical examinations made available pursuant to paragraph (j)(3)(i)(B) and (C) of this section shall be determined by an examining physician and, if requested by an employee, shall include pregnancy testing or laboratory evaluation of male fertility. Medical examinations made available pursuant to paragraph (j)(3)(i)(A) of this section shall include the following elements:

- A detailed work history and a medical history, with particular attention to past lead exposure (occupational and nonoccupational), personal habits (smoking, hygiene), and past gastrointestinal, hematologic, renal, cardiovascular, reproductive, and neurological problems
- A thorough physical examination, with particular attention to teeth, gums, hematologic, gastrointestinal, renal, cardiovascular, and neurological systems; pulmonary status should be evaluated if respiratory protection will be used
- A blood pressure measurement
- A blood sample and analysis to determine blood-lead level

- Hemoglobin and hematocrit determinations, red cell indices, and examination of peripheral smear morphology, zinc protoporphyrin, blood urea nitrogen, and serum creatinine
- A routine urinalysis with microscopic examination and any laboratory or other test that is relevant to lead exposure that the examining physician deems necessary by sound medical practice

Multiple Physician Review Mechanism

If the employer selects the initial physician who conducts any medical examination or consultation provided to an employee under this section, the employee may designate a second physician to review any findings, determinations, or recommendations of the initial physician; and to conduct such examinations, consultations, and laboratory tests as the second physician deems necessary to facilitate this review.

The employer shall promptly notify an employee of the right to seek a second medical opinion after each occasion for which an initial physician conducts a medical examination or consultation pursuant to this section. The employer may condition its participation in, and payment for, the multiple physician review mechanism on the employee doing the following within 15 days after receipt of the foregoing notification, or receipt of the initial physician's written opinion, whichever is later:

- The employee informing the employer that he or she intends to seek a second medical opinion.
- The employee initiating steps to make an appointment with a second physician
- If the findings, determinations, or recommendations of the second physician differ from those of the initial physician, then the employer and the employee shall ensure that efforts are made for the two physicians to resolve any disagreement

If the two physicians have been unable to quickly resolve their disagreement, then the employer and the employee through their respective physicians shall designate a third physician to review any findings, determinations, or recommendations of the prior physicians; and to conduct such examinations, consultations, laboratory tests, and discussions with the prior physicians as the third physician deems necessary to resolve the disagreement of the prior physicians.

The employer shall act consistent with the findings, determinations, and recommendations of the third physician, unless the employer and the employee reach an agreement that is otherwise consistent with the recommendations of at least one of the three physicians.

Information Provided to Examining and Consulting Physicians

The employer shall provide an initial physician conducting a medical examination or consultation under this section with the following information:

- A copy of this regulation for lead including all appendices
- A description of the affected employee's duties as they relate to the employee's exposure
- The employee's exposure level or anticipated exposure level to lead and to any other toxic substance (if applicable)
- A description of any personal protective equipment used or to be used
- Prior blood lead determinations
- All prior written medical opinions concerning the employee in the employer's possession or control

The employer shall provide the foregoing information to a second or third physician conducting a medical examination or consultation under this section on request either by the second or third physician, or by the employee.

Written Medical Opinions

The employer shall obtain and furnish the employee with a copy of a written medical opinion from each examining or consulting physician, which contains only the following information:

- The physician's opinion as to whether the employee has any detected medical condition that would place the employee at increased risk of material impairment of the employee's health from exposure to lead.
- Any recommended special protective measures to be provided to the employee, or limitations to be placed on the employee's exposure to lead.

- Any recommended limitation on employee's use of respirators, including a determination of whether the employee can wear a powered air purifying respirator if a physician determines that the employee cannot wear a negative pressure respirator.
- The results of the blood lead determinations.
- The employer shall instruct each examining and consulting physician to not reveal either in the written opinion or orally, or in any other means of communication with the employer, findings, including laboratory results, or diagnoses unrelated to an employee's occupational exposure to lead.
- Advise the employee of any medical condition, occupational or nonoccupational, that dictates further medical examination or treatment.

Alternate Physician Determination Mechanisms

The employer and an employee or authorized employee representative may agree on the use of any alternate physician determination mechanism in lieu of the multiple physician review mechanism provided by paragraph (j)(3)(iii) of this section so long as the alternate mechanism is as expeditious and protective as the requirements contained in this paragraph.

The employer shall ensure that any person who the company retains, employs, supervises, or controls does not engage in prophylactic chelation of any employee at any time. If therapeutic or diagnostic chelation is to be performed by any person in paragraph (j)(4)(i) of this section, the employer shall ensure that it be done under the supervision of a licensed physician in a clinical setting with thorough and appropriate medical monitoring and that the employee is notified in writing prior to its occurrence.

The employer shall remove an employee from work having an exposure to lead at or above the action level on each occasion that a periodic and a follow-up blood sampling test conducted pursuant to this section indicate that the employee's blood-lead level is at or above 50 µg/dL.

The employer shall remove an employee from work having an exposure to lead at or above the action level on each occasion that a final medical determination results in a medical finding, determination, or opinion that the employee has a detected medical condition that places the employee at increased risk of material impairment to health from exposure to lead.

For the purposes of this section, the phrase "final medical determination" means the written medical opinion about the employee's health status by the examining physician or, whenever relevant, the outcome of the multiple physician review mechanism or alternate medical determination mechanism used pursuant to the medical surveillance provisions of this section.

Where a final medical determination results in any recommended special protective measures for an employee, or limitations on an employee's exposure to lead, the employer shall implement and act consistent with the recommendation.

Return of the Employee to Former Job Status
The employer shall return an employee to his or her former job status:

- For an employee removed due to a blood-lead level at or above 50 μg/dL when two consecutive blood sampling tests indicate that the employee's blood-lead level is at or below 40 μg/dL
- For an employee removed due to a final medical determination, when a subsequent final medical determination results in a medical finding, determination, or opinion that the employee no longer has a detected medical condition that places the employee at increased risk of material impairment to health from exposure to lead

For this section's purposes, the requirement that an employer return an employee to his or her former job status is not intended to expand on or restrict any rights an employee has or would have had, absent temporary medical removal, to a specific job classification or position under the terms of a collective bargaining agreement.

Removal of Other Employee Special Protective Measure or Limitations
The employer shall remove any limitations placed on an employee or end any special protective measures provided to an employee pursuant to a final medical determination when a subsequent final medical determination indicates that the limitations or special protective measures are no longer necessary.

Employer Options Pending a Final Medical Determination

Where the multiple physician review mechanism, or alternate medical determination mechanism used pursuant to the medical surveillance provisions of this section, has not yet resulted in a final medical determination with respect to an employee, the employer shall act as follows:

- The employer may remove the employee from exposure to lead, provide special protective measures to the employee, or place limitations on the employee, consistent with the medical findings, determinations, or recommendations of any of the physicians who have reviewed the employee's health status.
- The employer may return the employee to his or her former job status, end any special protective measures provided to the employee, and remove any limitations placed on the employee, consistent with the medical findings, determinations, or recommendations of any of the physicians who have reviewed the employee's health status, with two exceptions.
  - If the initial removal, special protection, or limitation of the employee resulted from a final medical determination which differed from the findings, determinations, or recommendations of the initial physician
  - If the employee has been on removal status for the preceding 18 months due to an elevated blood-lead level, then the employer shall await a final medical determination

Provision of Medical Removal Protection Benefits

The employer shall provide an employee up to 18 months of medical removal protection benefits on each occasion that an employee is removed from exposure to lead or otherwise limited pursuant to this section.

For the purposes of this section, the requirement that an employer provide medical removal protection benefits means that, as long as the job the employee was removed from continues, the employer shall maintain the total normal earnings, seniority, and other employment rights and benefits of an employee, including the employee's right to his or her former job status as though the employee had not been medically removed from the employee's job or otherwise medically limited.

During the period of time that an employee is medically removed from his or her job or otherwise medically limited, the employer may condition the provision of medical removal protection benefits on the employee's participation in follow-up medical surveillance made available pursuant to this section.

If a removed employee files a claim for workers' compensation payments for a lead-related disability, then the employer shall continue to provide medical removal protection benefits pending disposition of the claim. To the extent that an award is made to the employee for earnings lost during the period of removal, the employer's medical removal protection obligation shall be reduced by such amount. The employer shall receive no credit for workers' compensation payments received by the employee for treatment-related expenses.

The employer's obligation to provide medical removal protection benefits to a removed employee shall be reduced to the extent that the employee receives compensation for earnings lost during the period of removal either from a publicly or employer-funded compensation program, or receives income from employment with another employer made possible by virtue of the employee's removal.

Voluntary Removal or Restriction of an Employee

Where an employer, although not required by this section to do so, removes an employee from exposure to lead or otherwise places limitations on an employee due to the effects of lead exposure on the employee's medical condition, the employer shall provide medical removal protection benefits to the employee equal to that required by paragraph (k)(2)(i) and (ii) of this section.

▶ EMPLOYEE INFORMATION AND TRAINING

The employer shall communicate information concerning lead hazards according to the requirements of OSHA's Hazard Communication Standard for the construction industry, 29 CFR 1926.59, including but not limited to the requirements concerning warning signs and labels, material safety data sheets (MSDS), and employee information and training. In addition, employers shall comply with the following requirements:

• The employer shall train each employee who is subject to exposure to lead at or above the action level on any day, or who is subject to exposure to lead compounds which may cause skin or eye irritation (e.g., lead arsenate, lead azide), in accordance with the requirements of this section. The employer shall institute a training program and ensure employee participation in the program.

- The employer shall provide the training program as initial training prior to the time of job assignment or prior to the start-up date for this requirement, whichever comes last.
- The employer shall also provide the training program at least annually for each employee who is subject to lead exposure at or above the action level on any day.
- The employer shall ensure that each employee is trained in the following:
  - The content of this standard and its appendices
  - The specific nature of the operations that could result in exposure to lead above the action level
  - The purpose, proper selection, fitting, use, and limitations of respirators
  - The purpose and a description of the medical surveillance program, and the medical removal protection program including information concerning the adverse health effects associated with excessive exposure to lead (with particular attention to the adverse reproductive effects on both males and females and hazards to the fetus and additional precautions for employees who are pregnant)
  - The engineering controls and work practices associated with the employee's job assignment including training of employees to follow relevant good work practices
  - The contents of any compliance plan in effect; instructions to employees that chelating agents should not routinely be used to remove lead from their bodies and should not be used at all except under the direction of a licensed physician; and the employee's right of access to records under 29 CFR 1910.20

Access to Information and Training Materials

At a minimum, the employee should have access to information and training materials, as follows:

- The employer shall make readily available to all affected employees a copy of this standard and its appendices.

- The employer shall provide, on request, all materials relating to the employee information and training program to affected employees and their designated representatives, and to the Assistant Secretary and the Director.
- The employer may use signs required by other statutes, regulations, or ordinances in addition to, or in combination with, signs required by this paragraph.
- The employer shall ensure that no statement appears on or near any sign required by this paragraph that contradicts or detracts from the meaning of the required sign.
- The employer shall post a warning sign, such as the following, in each work area where employees may have exposure to lead that is above the PEL.
- The employer shall ensure that signs required by this paragraph are illuminated and cleaned as necessary so that the legend is readily visible.

WARNING
LEAD WORK AREA
POISON
NO SMOKING OR EATING

▶ RECORDKEEPING

Exposure monitoring records shall include:

- The employer shall establish and maintain an accurate record of all monitoring and other data used in conducting employee exposure assessments as required in this section's paragraph (d).
- The date(s), number, duration, location, and results of each of the samples taken if any, including a description of the sampling procedure used to determine representative employee exposure where applicable
- A description of the sampling and analytical methods used and evidence of their accuracy
- The type of respiratory protective devices worn, if any
- Name, social security number, and job classification of the employee monitored and of all other employees whose exposure the measurement is intended to represent
- The environmental variables that could affect the measurement of employee exposure

The employer shall maintain monitoring and other exposure assessment records in accordance with the provisions of 29 CFR 1926.33.

Medical Surveillance

The employer shall establish and maintain an accurate record for each employee subject to medical surveillance as required by paragraph (j) of this section. This record shall include:

- The name, social security number, and description of the duties of the employee
- A copy of the physician's written opinions
- Results of any airborne exposure monitoring done on or for that employee and provided to the physician
- Any employee medical complaints related to exposure to lead

The employer shall keep, or ensure that the examining physician keeps, the following medical records:

- A copy of the medical examination results including medical and work history required under this section's paragraph (j)
- A description of the laboratory procedures and a copy of any standards or guidelines used to interpret the test results or references to that information
- A copy of the results of biological monitoring
- The employer shall maintain or ensure that the physician maintains medical records in accordance with the provisions of 29 CFR 1926.33

Medical Removals

The employer shall establish and maintain an accurate record for each employee removed from current exposure to lead pursuant to paragraph (k) of this section. Each record shall include:

- The name and social security number of the employee
- The date of each occasion that the employee was removed from current exposure to lead as well as the corresponding date on which the employee was returned to his or her former job status
- A brief explanation of how each removal was or is being accomplished

- A statement with respect to each removal indicating whether the reason for the removal was an elevated blood-lead level

The employer shall maintain each medical removal record for at least the duration of an employee's employment.

Objective Data

For purposes of this section, objective data are information demonstrating that a particular product or material containing lead or a specific process, operation, or activity involving lead cannot release dust or fumes in concentrations at or above the action level under any expected conditions of use. Objective data can be obtained from an industry-wide study or from laboratory product test results from manufacturers of lead-containing products or materials. The data the employer uses from an industry-wide survey must be obtained under workplace conditions closely resembling the processes, types of material, control methods, work practices, and environmental conditions in the employer's current operations.

The employer shall maintain the record of the objective data relied on for at least 30 years. The employer shall make available, on request, all records required to be maintained by paragraph (n) of this section to affected employees, former employees, and their designated representatives, and to the Assistant Secretary and the Director for examination and copying.

Whenever the employer ceases to do business, the successor employer shall receive and retain all records required to be maintained by paragraph (n) of this section. Whenever the employer ceases to do business and there is no successor employer to receive and retain the records required to be maintained by this section for the prescribed period, these records shall be transmitted to the Director.

At the expiration of the retention period for the records required to be maintained by this section, the employer shall notify the Director at least 3 months prior to the disposal of such records and shall transmit those records to the Director if requested within the period.

The employer shall also comply with any additional requirements involving transfer of records as set forth in 29 CFR 1926.33(h).

Observation of Monitoring

The employer shall provide affected employees or their designated representatives an opportunity to observe any monitoring of employee exposure to lead conducted pursuant to paragraph (d) of this section.

Whenever observation of the monitoring of employee exposure to lead requires entry into an area where the use of respirators, protective clothing, or equipment is required, the employer shall provide the observer with and ensure the use of such respirators, clothing, and equipment, and shall require the observer to comply with all other applicable safety and health procedures. Without interfering with the monitoring, observers shall be entitled to:

- Receive an explanation of the measurement procedures
- Observe all steps related to the monitoring of lead performed at the place of exposure
- Record the results obtained or receive copies of the results when returned by the laboratory

As you can see, OSHA sets forth plenty of rules for contractors to follow. In addition to OSHA, you have to work within the framework of EPA regulations and any state and local requirements. Large companies can afford safety officers to adhere to the rules. Most contractors do not have this luxury. Therefore, the burden may fall on your shoulders. If it does, take it seriously.

Now, let's move to the next chapter and review what the EPA requires when working with lead.

EPA Lead Renovation, Repair, and Painting Program Overview

5

In 2008, the Environmental Protection Agency enacted a new rule to deal with risks associated with lead. The Lead Renovation, Repair, and Painting Program (LRRPR) is huge. People affected by the rule created a lot of buzz about it. While OSHA regulations for working with lead are modest, EPA rulings more than make up for them. Digesting the volume of this key rule can take days. It is not a rule that one memorizes.

This chapter takes a look at this rule. Just what does this rule involve? Well, you are about to find out. The rule itself is so large that it covers about 175 pages of single-spaced text. Obviously you will not be working with all aspects of the ruling, but you will have to be very aware of all the elements that apply to you. It might help to make some notes as you go through this information.

You may be potentially affected by this action if you perform renovations of target housing or child-occupied facilities for compensation or dust sampling. *Target housing* is defined in section 401 of the Toxic Substances Control Act (TSCA) as any housing constructed prior to 1978, except housing for the elderly or persons with disabilities (unless any child under age 6 resides or is expected to reside in such housing) or any 0-bedroom dwelling.

Under this rule, a child-occupied facility is a building, or a portion of a building, constructed prior to 1978, visited regularly by the same child, under 6 years of age, on at least 2 different days within any week (Sunday through Saturday period), provided that each day's visit lasts at least 3 hours and the combined weekly visits last at least 6 hours, and the combined annual visits last at least 60 hours. Child-occupied facilities may be located in public or commercial buildings or in target housing.

Potentially affected entities may include, but are not limited to:

- Building construction (NAICS code 236)—for example, single-family housing construction, multifamily housing construction, residential remodelers
- Specialty trade contractors (NAICS code 238)—for example, plumbing, heating, and air-conditioning contractors, painting and wall covering contractors, electrical contractors, finish carpentry contractors, drywall and insulation contractors, siding contractors, tile and terrazzo contractors, and glass and glazing contractors
- Real estate (NAICS code 531)—for example, lessors of residential buildings and dwellings, residential property managers
- Child daycare services (NAICS code 624410)
- Elementary and secondary schools (NAICS code 611110)—for example, elementary schools with kindergarten classrooms
- Other technical and trade schools (NAICS code 611519)—for example, training providers
- Engineering services (NAICS code 541330) and building inspection services (NAICS code 541350)—for example, dust sampling technicians

This listing is not intended to be exhaustive, but rather provides a guide for readers regarding entities likely to be affected by this action. Other types of entities not listed in this unit could also be affected.

▶ LEAD

What is lead? Lead is a soft, bluish metallic chemical element mined from rock and found in its natural state all over the world. Lead is virtually indestructible, is persistent, and has been known since antiquity for its adaptability in making various useful items. In modern times, it has been used to manufacture many different products, including paint, batteries, pipes, solder, pottery, and gasoline. Through the 1940s, paint manufacturers frequently used lead as a primary ingredient in many oil-based interior and exterior house paints. Usage gradually decreased through the 1950s and 1960s as titanium dioxide replaced lead and as latex paints became more widely available.

Lead has been demonstrated to exert "a broad array of deleterious effects on multiple organ systems via widely diverse mechanisms of action." This array of health effects, the evidence for which is comprehensively described in EPA's *Air Quality Criteria for Lead* document, includes heme biosynthesis and related functions; neurological development and function; reproduction and physical development; kidney function; cardiovascular function; and immune function. There is also some evidence of lead carcinogenicity, primarily from animal studies, together with limited human evidence of suggestive associations.

Of particular interest for present purposes is the delineation of lowest observed effect levels for those lead-induced effects that are most clearly associated with blood lead less 10 µg/dL in children and/or adults and are, therefore, of greatest public health concern. As evident from the Criteria Document, neurotoxic effects in children and cardiovascular effects in adults are among those best substantiated as occurring at blood-lead concentrations as low as 5 to 10 µg/dL (or possibly lower); and these categories of effects are currently clearly of greatest public health concern. Other newly demonstrated immune and renal system effects among general population groups are also emerging as low-level lead-exposure effects of potential public health concern.

The overall weight of the available evidence provides clear substantiation of neurocognitive decrements being associated in young children with blood-lead concentrations in the range of 5 to 10 micrograms per deciliter (µg/dL), and possibly somewhat lower. Some newly available analyses appear to show lead effects on the intellectual attainment of preschool and school age children at population mean concurrent blood-lead levels ranging down to as low as 2 to 8 µg/dL. A decline of 6.2 points in full scale IQ for an increase in concurrent blood-lead levels from 1 to 10 µg/dL has been estimated, based on a pooled analysis of results derived from seven well-conducted prospective epidemiologic studies.

Epidemiologic studies have consistently demonstrated associations between lead exposure and enhanced risk of deleterious cardiovascular outcomes, including increased blood pressure and incidence of hypertension. A meta-analysis of numerous studies estimates that a doubling of blood-lead level (e.g., from 5 to 10 µg/dL) is associated with ~1.0 mm Hg increase in systolic blood pressure and ~0.6 mm Hg increase in diastolic pressure. Both epidemiologic and toxicologic studies have shown that environmentally relevant levels of lead affect many different organ systems.

The nervous system has long been recognized as a target of lead toxicity, with the developing nervous system affected at lower exposures than the mature system. While blood-lead levels in U.S. children ages 1 to 5 years have decreased notably since the late 1970s, newer studies have investigated and reported associations of effects on the neurodevelopment of children at population mean concurrent blood-lead levels ranging down to as low as 2 to 8 µg/dL. Functional manifestations of lead neurotoxicity during childhood include sensory, motor, cognitive, and behavioral impacts. Investigating associations between lead exposure and behavior, mood, and social conduct of children has been an emerging area of research. Early studies indicated linkages between lower-level lead toxicity and behavioral problems (e.g., aggression, attentional problems, and hyperactivity) in children.

Effects of lead on neurobehavior have been reported with remarkable consistency across numerous studies of various designs, populations studied, and developmental assessment protocols. The negative impact of lead on IQ and other neurobehavioral outcomes persist in most recent studies following adjustment for numerous confounding factors including social class, quality of caregiving, and parental intelligence. Moreover, these effects appear to persist into adolescence and young adulthood.

Cognitive effects associated with lead exposures that have been observed in some studies include decrements in intelligence test results, such as the widely used IQ score, and in academic achievement as assessed by various standardized tests as well as by class ranking and graduation rates. Associations between lead exposure and academic achievement observed in the above-noted studies were significant even after adjusting for IQ, suggesting that lead-sensitive neuropsychological processing and learning factors not reflected by global intelligence indices might contribute to reduced performance on academic tasks.

Other cognitive effects observed in studies of children have included effects on attention, executive functions, language, memory, learning, and visuospatial processing with attention and executive function effects observed. The evidence for the role of lead in this suite of effects includes experimental animal findings. These animal toxicology findings provide strong biological plausibility in support of the concept that lead may impact one or more of these specific cognitive functions in humans.

Further, lead-induced deficits observed in animal and epidemiological studies, for the most part, have been found to be persistent in the absence of markedly reduced environmental exposures. It is additionally important to note that there may be long-term consequences of such deficits over a lifetime. Studies examining aspects of academic achievement related to lead exposure indicate the association of deficits in academic skills and performance, which in turn lead to enduring and important effects on objective parameters of success in real life.

Lead bioaccumulates, and is only slowly removed, with bone lead serving as a blood lead source for years after exposure and possibly serving as a significant source of exposure. Bone accounts for more than 90% of the total body burden of lead in adults and 70% in children. In comparison to adults, bone mineral turns over much more quickly in children as a result of growth. Changes in blood-lead concentration in children are thought to parallel more closely to changes in total body burden. Therefore, blood-lead concentration is often used in epidemiologic and toxicological studies as an index of exposure and body burden for children.

Paint that contains lead can pose a health threat through various routes of exposure. House dust is the most common exposure pathway through which children are exposed to lead-based paint hazards. Dust created during normal lead-based paint wear (especially around windows and doors) can create an invisible film over surfaces in a house. Children, particularly younger children, are at risk for high exposures of lead-based paint dust via hand-to-mouth exposure, and may also ingest lead-based paint chips from flaking paint on walls, windows, and doors. Lead from exterior house paint can flake off or leach into the soil around the outside of a home, contaminating children's play areas. Cleaning and renovation activities may actually increase the threat of lead-based paint exposure by dispersing lead dust particles in the air and over accessible household surfaces. In turn, both adults and children can receive hazardous exposures by inhaling the dust or by ingesting lead-based paint dust during hand-to-mouth activities.

Renovation

Renovation is defined as the modification of any existing structure, or portion of a structure, that results in the disturbance of painted surfaces. The regulations specifically exclude lead-based paint abatement projects as well as small projects that disturb 2 square feet or less of painted surface per component, emergency projects, and renovations affecting components that have been found to be free of lead-based paint, as that term is defined in the regulations, by a certified inspector or risk assessor. These regulations require the renovation firm to document compliance with the requirement to provide the owner and the occupant with the *Protect Your Family From Lead in Your Home* (PYF) pamphlet. TSCA section 404 also allows states to apply for, and receive authorization to administer, the TSCA section 406(b) requirements.

Danger Levels

EPA establishes dangerous levels of lead in paint, dust, and soil. These hazard standards define lead-based paint hazards in target housing and child-occupied facilities as paint-lead, dust-lead, and soil-lead hazards. A paint-lead hazard is defined as any damaged or deteriorated lead-based paint, any chewable lead-based painted surface with evidence of teeth marks, or any lead-based paint on a friction surface if lead dust levels underneath the friction surface exceed the dust-lead hazard standards.

A dust-lead hazard is surface dust that contains a mass-per-area concentration of lead equal to or exceeding 40 micrograms per square foot ($\mu g/ft^2$) on floors or $250\,\mu g/ft^2$ on interior windowsills based on wipe samples. A soil-lead hazard is bare soil that contains total lead equal to or exceeding 400 parts per million (ppm) in a play area or average of 1200 ppm of bare soil in the rest of the yard based on soil samples.

To establish the risk and work decisions for a job a risk assessment is needed. This is often done with field inspections and samples. The following are examples of exterior areas that are normally considered potential risks for lead paint:

- Fences
- Storage sheds and garages
- Laundry line posts
- Swing sets and other play equipment

When conducting field sampling, it is common to use standard forms to record findings. Different forms of rulings apply to single-family homes, rental units, and multifamily dwellings. Examples of these forms can be seen in Figures 5.1, 5.2, and 5.3.

Management Data for Risk Assessment of Lead-Based
Paint Hazards in Rental Dwellings (Optional)

NOTE: This form is designed for multiple rental dwellings under one ownership. Such dwellings may be in one
property or many.

Part 1: Identifying information

Name of property owner_____

Name of building or development (if applicable)_____

Number of dwelling units_____

Number of buildings_____

Number of individual dwelling units/building_____

Date of construction (if one property)_____(if between 1960–1978, consider a screen risk assessment)

Date of substantial rehab, if any_____

List of addresses of dwellings (attach list if more than 10 dwellings are present)

Street address, city, state	Dwelling unit no.	Year built (if known)	Number of children 0–6 years old	Recent code violation reported by owner?	Chronic maintenance problem reported by owner?

Record number and locations of common child play areas (onsite playground, backyards, etc.)

Number_____

_____   _____   _____
_____   _____   _____
_____   _____   _____

Figure 5.1 Management data for risk assessment of lead-based paint hazards in rental dwellings.

▶ FIELD SAMPLING STUDY

EPA conducted a study in the following four phases:

- Phase I, the Environmental Field Sampling Study, evaluated the amount of leaded dust released by the following activities:
  - Paint removal by abrasive sanding
  - Removal of large structures, including demolition of interior plaster walls
  - Window replacement
  - Carpet removal
  - HVAC repair or replacement, including duct work
  - Repairs resulting in isolated small surface disruptions, including drilling and sawing into wood and plaster

Part 2: Management Information

1. List names of individuals who have responsibility for lead-based paint. Include owner, property manager (if applicable), maintenance supervisor and staff (if applicable), and others. Include any training in lead hazard control work (by inspector, supervisor, worker, etc.) that has been completed. Use additional pages, if necessary.

This information will be needed to devise the risk management plan contained in the risk assessor's report.

Name	Position	Training completed (if none, enter "None")
	Owner	
	Property manager	
	Maintenance	

2. Have there been previous lead-based paint evaluations?

_____Yes_____No (If yes, attach the report)

3. Has there been previous lead hazard control activity?

_____Yes_____No (If yes, attach the report)

4. Maintenance usually conducted at time of dwelling turnover, including typical cleaning, repainting, and repair activity.

Repainting: _____
Cleaning: _____
Repair: _____
Other: _____
Comments: _____

5. Employee and worker safety plan

a. Is there an occupational safety and health plan for maintenance workers?

_____Yes_____No (If yes, attach plan)

b. Are workers trained in lead hazard recognition?

_____Yes_____No If yes, who performed the training?_____

Figure 5.1 *Cont'd*

- Phase II, the Worker Characterization and Blood-Lead Study, involved collecting data on blood lead and renovation and remodeling activities from workers.
- Phase III, the Wisconsin Childhood Blood-Lead Study, was a retrospective study focused on assessing the relationship between renovation and remodeling activities and children's blood-lead levels.
- Phase IV, the Worker Characterization and Blood-Lead Study of R&R Workers Who Specialize in Renovations of Old or Historic Homes, was similar to Phase II, but the study focused on individuals who worked primarily in old historic buildings.

    c. Are workers involved in a hazard communication program?

        _____Yes_____No

    d. Are workers trained in proper use of respirators?

        _____Yes_____No

    e. Is there a medical surveillance program?

        _____Yes_____No

6. Is a HEPA vacuum available?

    _____Yes_____No

7. Are there any onsite licensed or unlicensed day-care facilities?

    _____Yes_____No If yes, give location_____

8. Planning for resident children with elevated blood-lead levels

    a. Who would respond for the owner if a resident child with an elevated blood-lead level is identified?

      _____

    b. Is there a plan to relocate such children?

        _____Yes_____No If yes, where?_____

    c. Does the owner know if there ever has been a resident child with an elevated blood-lead level?

        _____Yes_____No_____Unknown

9. Owner Inspections

    a. Are there periodic inspections of all dwellings by the owner?

        _____Yes_____No If yes, how often?_____

    b. Is the paint condition assessed during these inspections?

        _____Yes_____No

10. Have any of the dwellings ever received a housing code violation notice?

    _____Yes_____No_____Unknown

    If yes, describe code violation_____

11. If previously detected, unabated lead-based paint exists in the dwelling, have the residents been informed?

    _____Yes_____No_____Not Applicable

Figure 5.1 *Cont'd*

Characterization of Dust Lead Levels after Renovation, Repair, and Painting Activities

EPA conducted a field study (Characterization of Dust Lead Levels after Renovation, Repair, and Painting Activities—the "Dust Study") to characterize dust lead levels resulting from various renovation, repair, and painting activities. This study, completed in January 2007, was designed to compare environmental lead levels at appropriate stages after various types of renovation, repair, and painting preparation activities were performed on the interiors and exteriors of target housing units and child-occupied facilities. All of the jobs disturbed more than 2 square feet of lead-based paint, so they would not have been eligible for the minor maintenance exception from the 2006 Proposal.

The renovation activities were conducted by local professional renovation firms, using personnel who received lead-safe work practices training using the curriculum developed by EPA and HUD,

Substrate Correction Values

Page 2 of 5

Address/Unit No. _918 Fenway Drive_

Oldtown, Maryland 21334

Date _August 15, 1997_   XRF Serial No. _RS-1967_

Inspector Name _Mo Smith_   Signature

Use this form when the *XRF Performance Characteristics Sheet* indicates that correction for substrate bias is needed.

	Substrate	Brick	Concrete	Drywall	Metal	Plaster	Wood
Location 1	First Reading				0.10		0.14
	Second Reading				0.09		0.13
	Third Reading				0.09		0.12
Location 2	First Reading				0.10		0.11
	Second Reading				0.09		0.12
	Third Reading				0.11		0.12
	Correction Value (Average of the Six Readings)				0.10		0.12

Transfer Correction Value for each substrate to the 'Correction Value' column of the LBP Testing Data Sheet.

Notes:

      Metal: Location 1 - Door frame, Side B, Room 2
                Location 2 - Door frame, Side C, Room 3

      Wood: Location 1 - Window Sill, Side A, Room 1
                Location 2 - Window Sill, Side B, Room 2

Completed Form 7.3

Figure 5.2 Sample of completed Substrate Correction Values form.

"Lead Safety for Remodeling, Repair, and Painting." The activities conducted represented a range of activities that would be permitted under the 2006 Proposal, including work practices that are restricted or prohibited for abatements under 40 CFR 745.227(e)(6). Of particular interest was the impact of using specific work practices that renovation firms would be required to use under the proposed rule, such as the use of plastic to contain the work area and a multistep cleaning protocol, as opposed to more typical work practices.

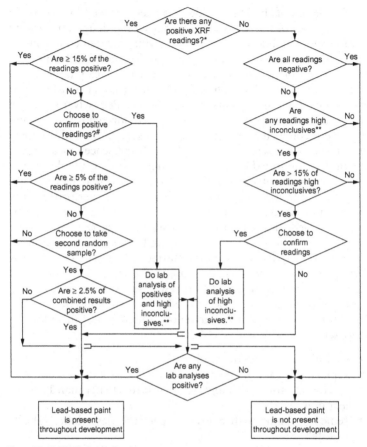

Figure 5.3 Multifamily decision flowchart.

*"Positive," "negative," and "inconclusive" X-Ray Fluorescence (XRF) readings are determined in accordance with the XRF Instruments' Performance Characteristics sheet as described in the *HUD Guidelines for the Evaluation and Control of Lead Hazards in Housing*.

**A high inconclusive reading is an XRF reading at or above the midpoint of the inconclusive range. For example, if the inconclusive range is 0.41 to 1.39, its midpoint (average) is 090; a reading ranging from 0.90 to 1.39 would be a high inconclusive reading.

#Any paint or coating may be assumed to be lead-based, even without XRF or laboratory analysis. Similiarly, any XRF reading may be confirmed by laboratory analysis.

The design of the Dust Study was peer-reviewed by experts in fields related to the study. They reviewed the design and quality assurance plan independently and provided written comments to EPA. In the Dust Study, 12 different interior and 12 different exterior renovation activities were performed at 7 vacant target housing units in Columbus, Ohio, and 8 vacant target housing units (including four apartments) in Pittsburgh, Pennsylvania. Three different interior and three different exterior renovation activities were conducted at a building representing a child-occupied facility, a vacant school in Columbus. The presence of lead-based paint was confirmed by laboratory analysis before a building was assigned a particular renovation activity or set of activities.

Before interior renovation activities were performed, the floors and windowsills in the work area and adjacent rooms were cleaned. In most cases, prework cleaning resulted in dust lead levels on floors of less than $10\,\mu g/ft^2$; nearly all floors were less than $40\,\mu g/ft^2$ before work started. Most windowsills that would be used for later sampling were cleaned to dust lead levels less than $250\,\mu g/ft^2$. In the few cases where that level was not achieved on a windowsill needed for sampling, dust collection trays were used. Interior renovation activities included the following jobs:

- Making cut-outs in the walls
- Replacing a window from the inside
- Removing paint with a high-temperature (greater than 1100°F) heat gun
- Removing paint with a low-temperature (less than 1100°F) heat gun
- Removing paint by dry scraping
- Removing kitchen cabinets
- Removing paint with a power planer

To illustrate the impact of the containment plastic and the specialized cleaning and cleaning verification protocol that would be required by the 2006 Proposal, each activity was performed a minimum of four times:

- With the plastic containment described in the 2006 Proposal followed by the cleaning protocol described in the proposal
- With the plastic containment described in the 2006 Proposal followed by dry sweeping and vacuuming with a shop vacuum

- With no plastic containment followed by the cleaning protocol described in the 2006 Proposal
- With no plastic containment followed by dry sweeping and vacuuming with a shop vacuum

Dust samples were collected after the renovation work was completed, after cleaning, and after cleaning verification. If a building was being used again for the same job under different work practices, or for a completely different job, the unit was recleaned and retested prior to starting the next job. All buildings were cleaned and tested after the last job.

Geometric mean post-work, precleaning floor dust lead levels in the work room were as follows (in $\mu g/ft^2$):

- Cut-outs: 422
- Kitchen cabinet removal: 958
- Low-temperature heat gun: 2080
- Dry scraping: 2686
- Window replacement: 3993
- High-temperature heat gun: 7737
- Power planing: 32,644

Power planing is an activity very similar to power sanding in which a machine that operates at high speed generating large quantities of dust is used. Where baseline practices (i.e., no containment, dry sweeping, and vacuuming with a shop vacuum) were used, the geometric mean post-job floor dust lead levels in the work room were as follows (in $\mu g/ft^2$):

- Cut-outs: 22
- Kitchen cabinet removal: 58
- Low-temperature heat gun: 41
- Dry scraping: 66
- Window replacement: 135
- High-temperature heat gun: 445
- Power planing: 450

The package of proposed rule requirements (i.e., containment, specialized cleaning, and cleaning verification) resulted in the lowest geometric mean dust lead levels in the work room at the end of a job. These results were as follows (in $\mu g/ft^2$):

- Cut-outs: 5
- Kitchen cabinet removal: 12

- Low-temperature heat gun: 24
- Dry scraping: 30
- Window replacement: 33
- High-temperature heat gun: 36
- Power planing: 148

Windowsill sample results were similar; the geometric mean dust lead levels after renovation activities performed in accordance with the proposed rule exceeded $250\,\mu g/ft^2$ only where power planing or a high-temperature heat gun were used. When baseline practices were used, the geometric mean dust lead levels on the windowsills exceeded $250\,\mu g/ft^2$ for kitchen cabinet removal, window replacement, high-temperature heat gun use, and power planing.

▶ EXTERIOR RENOVATION ACTIVITIES

Exterior renovation activities performed as part of the study included the following:

- Replacing a door and doorway
- Replacing fascia boards, soffits, and other trim
- Removing paint with a high-temperature (greater than 1100°F) heat gun
- Removing paint with a low-temperature (less than 1100°F) heat gun
- Removing paint by dry scraping
- Removing paint with a needle gun
- Removing paint with power sanding or grinding
- Removing paint with a torch or open flame

For the exterior jobs, plastic sheeting was placed on the ground to catch the debris and dust from the job, in accordance with the requirements of the proposed rule. Additional plastic sheeting was laid out beneath and beyond the "proposed rule" plastic. Trays to collect dust and debris were placed on top of and underneath the proposed rule plastic. Trays were also placed just outside of the proposed rule plastic to assess how far the dust was spreading. A vertical containment, as high as the work zone, was erected at the end of the additional plastic.

The use of the proposed rule plastic as a ground covering captured large amounts of leaded dust. For all job types except removing paint with a torch, there was a substantial difference between the amount of lead captured by the proposed rule plastic and the amount under the proposed rule plastic. Including both bulk debris and dust, geometric mean lead levels in exterior samples from the collection trays on top of the proposed rule plastic ranged from a low of $60,662\,\mu g/ft^2$ for the door replacement activity to a high of $7,216,358\,\mu g/ft^2$ for removing paint with a high-temperature heat gun. Geometric mean lead levels from the collection trays under the proposed rule plastic ranged from a low of $32\,\mu g/ft^2$ for door replacement to $8565\,\mu g/ft^2$ for removing paint with a torch.

This regulatory action was supported by the Dust Study just discussed. Therefore, EPA conducted a peer review in accordance with the *Final Information Quality Bulletin for Peer Review* of the U.S. Office of Management and Budget (OMB). EPA requested this review from the Clean Air Scientific Advisory Committee (CASAC) Lead Review Panel. The CASAC, which is comprised of seven members appointed by the EPA Administrator, was established under the Clean Air Act as an independent scientific advisory committee. The CASAC's comments on the Dust Study, along with EPA's responses, have been placed into the public docket for this action. More information on the CASAC consultation process, along with background documents, is available on EPA's website at *www.epa.gov/lead/pubs/casac.htm*.

The Dust Study was reasonably well designed, considering the complexity of the problem, and that the report provided information not available from any other source. The study indicated that the rule cleaning procedures reduced the residual lead (Pb) remaining after a renovation more than did the baseline cleaning procedures. Another positive aspect of the Dust Study was that it described deviations from the protocol when they occurred.

The CASAC panel also contended that the limited data from residential housing units and child-occupied facilities included in the Dust Study most likely do not represent a statistically valid sample of housing at the national level. It noted that there are aspects of the study that would underestimate the levels of lead loadings while other aspects of the study would overestimate the loadings. EPA agrees that the Dust Study is not nationally representative of all housing. EPA notes that there are several reasons why this is the case, including the fact that all of the housing studied was built during 1925 or earlier, and a large number of the floors were in poor condition.

A major purpose of the Dust Study was to assess the proposed work practices. A statistically valid sample of housing at the national level is not needed to assess the work practices. If anything, the Dust Study is conservative with respect to the age of housing because it studied older houses and therefore is appropriate for assessing the effectiveness of the work practices.

In addition to the Dust Study, which directly supported this regulatory action, several other studies are discussed that may or may not have been peer reviewed.

Lead-Safe Work Practices Survey Project

The National Association of Home Builders (NAHB) conducted a survey that assessed renovation and remodeling activities to measure levels of lead dust generated by home improvement contractors. The stated objective of this survey, completed in November 2006, was to measure the amount of lead dust generated during typical renovation and remodeling activities and assess whether routine renovation and remodeling activities increased lead dust levels in the work area and on the property.

The activities evaluated during the survey were selected in consultation with remodeling contractors. NAHB believes that these activities represent the most common jobs performed by renovation and remodeling firms. The renovations were performed by professional renovation and remodeling contractors from each of the communities where the properties were located. All of the workers who participated in this project had previously attended and successfully completed the EPA/HUD curriculum for "Lead Safety for Remodeling, Repair, and Painting."

According to the NAHB survey, an EPA-certified lead-based paint inspector confirmed the presence of lead-based paint in all of the properties considered for this survey. Previous inspection reports were consulted if the inspections conformed to the HUD guidelines for lead-based paint inspections. Properties used in this survey included a single-family home in Illinois, two single-family homes and a duplex in Connecticut, and an apartment above a storefront in Wisconsin.

The NAHB survey evaluated the following activities:

- Wall and ceiling removal (demolition)
- Wall and ceiling modification
- Window and door removal and/or replacement (no sanding)

- Window and door alteration (no sanding)
- Sanding on windows and doors
- Kitchen or bath cabinet removal
- Baseboard and stair removal
- Surface preparation (sanding)
- Sawing into wood and plaster

Activities were performed in one of three ways: using the work practices presented in the EPA/HUD curriculum, using modified work practices (one or more of the dust control or cleanup methods discussed in the EPA/HUD curriculum), or routine renovation practices.

Area air samples were collected before, during, and after the work activity. Personal breathing zone air samples were collected during the work activity. Dust wipe samples were collected before work started and after final clean-up. Dust wipe samples were routinely collected from floors near the work activity and in some cases collected from a windowsill and/or window well.

In comparing the mean dust lead levels before the activities with the mean dust lead levels after the activities, the NAHB concluded that the renovation activities surveyed did not create new lead dust hazards overall. However, even after clean-up was conducted, over half of the 60 individual renovation activities studied resulted in an increase in dust lead levels on at least one surface. In most cases, the increase was considerably greater than the regulatory dust-lead hazard standard for that surface.

In the 2006 Proposal, EPA proposed to conclude that any renovation activity that disturbs lead-based paint can create significant amounts of leaded dust, that most activities created lead-based paint hazards, and that some activities can be reasonably anticipated to create lead-based paint hazards. EPA's proposed conclusions were based upon the results of the Environmental Field Sampling Study, which examined, on a variety of components using a variety of tools and methods, activities that EPA had determined were representative of the paint-disturbing activities that typically occur during renovations. The activities were:

- Paint removal by abrasive sanding
- Window replacement
- HVAC duct work
- Demolition of interior plaster walls

- Drilling into wood
- Drilling into plaster
- Sawing into wood
- Sawing into plaster

Specifically, EPA proposed to conclude that all of the activities studied in the Environmental Field Sampling Study, with the exception of drilling into plaster, can create lead-based paint hazards. With respect to drilling into plaster, where lead-based paint is present, EPA proposed to conclude that this activity can reasonably be anticipated to create lead-based paint hazards. The Environmental Field Sampling Study found that, with the exception of drilling into plaster, all renovation and remodeling activities, when conducted where lead-based paint is present, generated lead loadings on floors at a distance of 5 to 6 feet from the activity that exceeded EPA's dust-lead hazard standard of $40 \mu g/ft^2$. However, upon further review, it is apparent that the study also found that drilling into plaster created dust lead levels in the immediate vicinity of the activity that exceeded the dust-lead hazard standard. Thus, all the activities studied did in fact create lead-based paint hazards. The hazards can be reduced with the use of proper tools.

The 2006 Proposal cited the other phases of the TSCA section 402(c)(2) renovation and remodeling study to support EPA's proposed determination that any renovation, remodeling, or painting activity that disturbs lead-based paint can be reasonably anticipated to create lead-based paint hazards. Phase III, the Wisconsin Childhood Blood-Lead Study, found that children who live in homes where renovation and remodeling activities were performed within the past year are 30% more likely to have a blood-lead level that equals or exceeds $10 \mu g/dL$, the level of concern established by CDC, than children living in homes where no such activity has taken place recently.

Phases II and IV of the study, which evaluated worker exposures from renovation and remodeling activities, provide additional documentation of the significant and direct relationship between blood-lead levels and the conduct of certain renovation and remodeling activities. Phase II found a statistically significant association between increased blood-lead levels and the number of days spent performing general renovation and remodeling activities, paint removal, and cleanup in pre-1950 buildings in the past month. Phase IV of the study found that persons performing renovation and remodeling activities in old historic buildings are more likely to have elevated blood-lead levels than persons in the general population of renovation and remodeling workers.

In light of EPA's proposed determination, the 2006 Proposal included revisions to the existing Lead-Based Paint Activities Regulations to extend them to renovation, remodeling, and painting activities in target housing, with certain exceptions. In proposing to extend these regulations to renovation, remodeling, and painting activities in child-occupied facilities, the 2007 Supplemental Proposal incorporated the proposed TSCA section 402(c)(3) determination.

Since the 2006 Proposal, EPA conducted the Dust Study and NAHB submitted the results of its survey. The results of the Dust Study confirm that renovation and remodeling activities that disturb lead-based paint create lead-based paint hazards. The Dust Study evaluated a number of common renovation activities, including replacing windows, removing kitchen cabinets, cutting into walls, and removing paint by high- and low-temperature heat guns, power tools, and dry scraping. The geometric mean post-work dust lead levels on work room floors ranged from a low of $422\,\mu g/ft^2$, or 10 times the dust-lead hazard standard for floors, for cut-outs, to a high of $32,644\,\mu g/ft^2$ for power planing.

Thus, all of the activities evaluated in the Dust Study created floor dust lead levels that exceeded $40\,\mu g/ft^2$, one of the measures that, in 40 CFR 745.65, defines a lead-based paint hazard. It is more difficult to evaluate the effect of disturbing lead-based paint in the NAHB Survey, since the survey did not involve collecting samples after work had been performed but before the post-renovation cleaning had begun. Nevertheless, even after post-renovation cleaning using a variety of methods, in more than half of the 60 experiments performed in this survey, the post-cleaning dust wipe sample results for at least one surface showed an increase greater than the TSCA section 403 hazard standard over prework levels. These experiments showing increased dust lead levels cover the range of activities evaluated in the NAHB Survey. Table 5.1 shows data on single-family homes based on the year of construction.

Therefore, in this action, EPA is issuing its determination that renovation, repair, and painting activities that disturb

TABLE 5.1 Privately Owned Dwellings with Lead-Based Paint by Age and Amount

YEAR OF CONSTRUCTION	TOTAL OCCUPIED UNITS[1]	PERCENT WITH LEAD-BASED PAINT[2]	AVERAGE AREA WITH LEAD-BASED PAINT ($\geq$1 MG/CM²) ON INTERIOR AND EXTERIOR SURFACES (FT²)[3]
1960–1979	35,681,000	62	466
1940–1959	20,476,000	80	1,090
Before 1940	21,018,000	90	1,996

[1]Total units data are from the 1987 American Housing Survey.
[2]The approximate 95% confidence intervals for the estimated percentages are:
    1960–1979 and before 1940 equals ±10%; 1940–1959 equals ±9%.
[3]Calculated from Tables 3-14 and 3-15 of the source. Average is calculated using only
    units with lead-based paint.
Source: *Comprehensive and Workable Plan for the Abatement of Lead-Based Paint in
    Privately Owned Housing: A Report to Congress* (HUD, 1990).

lead-based paint create lead-based paint hazards. Because the evidence shows that all such activities in the presence of lead-based paint create lead-based paint hazards, EPA is modifying its proposed finding, which distinguished between activities that create lead-based paint hazards and those that can reasonably be anticipated to create lead-based paint hazards, and instead concludes that renovation activities that disturb lead-based paint create lead-based paint hazards. Indeed, no commenter submitted data indicating that any renovation, repair, or painting activity should be exempt from regulation because it does not create lead-based paint hazards.

EPA received a large number of comments on this proposed finding. Many expressed support for EPA's determination that any renovation, repair, or painting activity that disturbs lead-based paint creates lead-based paint hazards. Some commenters, while expressing their support for this determination, also opined that the regulatory dust-lead hazard standards for floors and windowsills are too high. These commenters argued that recent scientific evidence shows that children experience adverse health effects at lower blood-lead levels than previously thought, and since EPA's regulatory dust-lead hazard standards were set with reference to a blood-lead level of 10 µg/dL, the CDC level of concern, the dust-lead hazard standards must be lowered. EPA agrees that recent studies demonstrate that neurocognitive effects occur at blood-lead levels below the current CDC level of concern.

In fact, EPA's most recent "Air Quality Criteria for Lead" document, issued in October 2006, describes several epidemiologic studies published in the last 5 years that observed significant lead-induced IQ decrements in children with some effects observed at blood-lead levels of 5 μg/dL and lower. The document also notes that other recent studies observed significant associations at low blood-lead levels for other neurotoxicity endpoints in addition to IQ, such as arithmetic and reading scores, attentional behavior, and neuromotive function. However, EPA is not addressing the appropriateness of the existing dust-lead hazard standards in this rulemaking.

The original hazard standards were set through a separate rulemaking process under TSCA section 403 that allowed for input from all of the parties that would be affected by the standards. Furthermore, EPA is concerned that a full review of the available evidence and other considerations affecting the hazard standards as part of this rulemaking would result in a significant delay in promulgating training, certification, and work practice standards for renovation activities. EPA did not propose to modify the TSCA section 403 hazard standard levels in this rulemaking and has not undertaken the significant analyses that would need to be performed to establish different standards. Accordingly, EPA is not able, in this final rule, to modify the regulatory hazard standard. In any event, since EPA finds that renovation activities that disturb lead-based paint create lead-paint hazards, lowering the hazard standard would not affect EPA's finding.

Some commenters objected to EPA's proposed determination that renovation, repair, or painting activities that disturb lead-based paint create lead-based paint hazards. Some commenters interpreted EPA's statutory authority to regulate renovation and remodeling under TSCA section 402(c)(3) as being limited to those renovation and remodeling activities for which EPA can prove a link between the activity and the blood-lead action level established by CDC for public health intervention.

These commenters contend that the failure to prove such a link means that renovation and remodeling activities do not create lead-based paint hazards. This interpretation is not supported by the plain language of the statute. TSCA section 402(c) (3) requires EPA to regulate renovation and remodeling activities that create lead-based paint hazards. The term *lead-based paint hazard* is defined in TSCA section 401 as "any condition that causes exposure to lead from lead-contaminated dust . . . that would result in adverse human health effects as established by the Administrator under this subchapter." TSCA section 403 directs EPA to promulgate regulations which "identify, for purposes of this subchapter and the Residential Lead-Based Paint Hazard Reduction Act of 1992, lead-based paint hazards, lead-contaminated dust, and lead-contaminated soil."

TSCA section 403 regulations define dust-lead hazards as levels that equal or exceed $40\,\mu g/ft^2$ of lead on floors or $250\,\mu g/ft^2$ of lead on interior windowsills. Therefore, EPA interprets TSCA as directing it to regulate renovation and remodeling activities if such activities create dust lead levels that exceed the standards for dust-lead hazards established under TSCA section 403. Again, the Environmental Field Sampling Study, the Dust Study, and the NAHB survey all demonstrate that renovation and remodeling activities that disturb lead-based paint create dust lead levels that exceed the hazard standards in 40 CFR 745.65. EPA also interprets the scientific evidence for a link between renovations and the CDC blood-lead action level differently than do these commenters.

EPA's Wisconsin Childhood Blood-Lead Study, described more fully in Unit III.C.1.c. of the preamble to the 2006 Proposal, provides ample evidence of a link between renovation activities and elevated blood-lead levels in resident children. This peer-reviewed study concluded that general residential renovation and remodeling is associated with an increased risk of elevated blood-lead levels in children and that specific renovation and remodeling activities are also associated with an increase in the risk of elevated blood-lead levels in children. In particular, removing paint (using open flame torches, using heat guns, using chemical paint removers, and wet scraping/sanding) and preparing surfaces by sanding or scraping significantly increased the risk of elevated blood-lead levels. Some of the commenters on this rule cited this as evidence that work performed by paid professional renovators does not create a statistically significant risk of an elevated blood-lead level in a resident child.

EPA agrees that the study report's Table 3-13, which presents the results of analyses using one of the sets of models used to interpret study data, indicates that, with respect to the persons performing the work, the only statistically significant result associated with increased risk of elevated blood-lead levels was work performed by a relative or friend not in the household. Work performed by professional renovators was associated with an increased risk of an elevated blood-lead level, but the association was not statistically significant. As explained more fully in a memorandum summarizing additional analyses of the data from this study, the table does not indicate that professional contractors were not responsible for creating lead exposure hazards. Rather, it indicates that renovation activities performed by professional contractors are no more or less hazardous than renovation activities performed by most of the other categories of persons identified in the survey responses collected as part of the study. It is also important to note that, while these commenters focus on a blood-lead level of $10 \, \mu g/dL$ as a threshold, this level is not and has not been considered by CDC or EPA as a threshold for adverse effects.

One commenter also dismissed the two studies from New York that EPA cited as supporting the findings of the Wisconsin Childhood Blood-Lead Study. In 1995, the New York State Department of Health assessed lead exposure among children resulting from home renovation and remodeling in 1993–1994. A review of the health department records of children with blood-lead levels equal to or greater than $20 \, \mu g/dL$ identified 320, or 6.9%, with elevated blood-lead levels that were attributable to renovation and remodeling. The commenter noted that this study suffered from a number of limitations, including the fact that it was not a case-control study; that is, the group of children with elevated blood-lead levels attributed to renovation and remodeling was not compared with a similar group of households that had not undergone renovation during the period. EPA agrees that this is an important limitation of this study. However, with respect to the other limitations noted by this commenter, the authors of the report felt that most of these limitations would likely result in an underestimation of the burden of lead exposure associated with renovation and remodeling. Table 5.2 shows hazard levels for lead-based paint risk assessments.

TABLE 5.2 Hazard Levels for Lead-Based Paint Risk Assessments

MEDIA	LEVEL	
Deteriorated paint (single-surface)	5,000 µg/g or 1 mg/cm^2	
Deteriorated paint (composite)	5,000 µg/g or 1 mg/cm^2	
	Number of subsamples	
Dust (wipe sampling only) (includes both single-surface and composite)	Risk assessment	Risk assessment screen (dwellings in good condition only)
Carpeted floors*	100 µg/ft^2	50 µg/ft^2
Hard floors*	100 µg/ft^2	50 µg/ft^2
Interior window sills	500 µg/ft^2	250 µg/ft^2
Window troughs	800 µg/ft^2	400 µg/ft^2
Bare soil (dwelling perimeter and yard)	2,000 µg/g	
Bare soil (small high-contact areas such as sandboxes and gardens)	400 µg/g	
Water (optional)—first draw	15 ppb (µg/dL)	

*Whenever possible, sample hard floors, not carpets.

The other study cited by EPA as supporting the Wisconsin Childhood Blood-Lead Study conclusions was a case-control study that assessed the association between elevated blood-lead levels in children younger than 5 years and renovation or repair activities in homes in New York City. EPA notes that the authors show that when dust and debris was reported (by respondents via telephone interviews) to be "everywhere" following a renovation, the blood-lead levels were significantly higher than in children at homes that did not report remodeling work. On the other hand, when the respondent reported either "no visible dust and debris" or that "dust and debris was limited to the work area," there was no statistically significant effect on blood-lead levels relative to homes that did not report remodeling work.

Although the study found only a weak and nonsignificant link between a report of any renovation activity and the likelihood that a resident child had an elevated blood-lead level, the link to the likelihood of an elevated blood-lead level was statistically significant for surface preparation by sanding and for renovation work that spreads dust and debris beyond the work area.

The researchers noted the consistency of their results with EPA's Wisconsin Childhood Blood-Lead Study. EPA notes that this confirms that keeping visible dust and debris contained to the work area is important for limiting children's exposures to lead dust, rather than providing substantial arguments for the effectiveness of visual inspection.

In sum, EPA's finding that renovation and remodeling activities create lead-based paint hazards is not dependent upon establishing a correlation between such activities and elevated blood-lead levels. Rather, it rests on the fact that, as demonstrated by EPA's Environmental Field Sampling Study, EPA's Dust Study, and by the NAHB survey, such activities create lead-based paint hazards as defined by EPA regulations. Moreover, EPA disagrees that there is no scientific support for establishing a relationship between elevated blood-lead levels in children and renovation activities. While EPA interprets these studies as supporting such a relationship and believes these studies further support its finding, it is not a determinative factor.

▶ THE FINAL RULE

How did the final rule come to exist? Here is the explanation. Given EPA's determination that renovation, repair, and painting activities that disturb lead-based paint create lead-based paint hazards, TSCA section 402(c)(3) directs EPA to revise the Lead-based Paint Activities Regulations to apply to these activities. EPA does not interpret its statutory mandate to require EPA to apply the existing TSCA section 402(a) regulations to renovations without change. By using the word "revise," and creating a separate subsection of the statute for renovation, the Environmental Protection Agency believes that Congress intended that EPA make revisions to those existing regulations to adapt them to a very different regulated community.

As discussed next, there are significant differences between renovations and abatements. Accordingly, this final rule does not merely expand the scope of the current abatement requirements to cover renovation and remodeling activities. Rather, EPA has carefully considered the elements of the existing abatement regulations and revised them as necessary to craft a rule that is practical for renovation, remodeling, and painting businesses and their customers, taking into account reliability, effectiveness, and safety as directed by TSCA section 402(a). Specifically, the agency concludes that the training, containment, cleaning, and cleaning verification requirements in this final rule achieve the goal of minimizing exposure to lead-based paint hazards created during renovation, remodeling, and painting activities, taking into account reliability, effectiveness, and safety.

In taking safety into account, EPA looked to the statutory directive to regulate renovation activities that create lead-based paint hazards. Although there is no known level of lead exposure that is safe, EPA does not believe the intent of Congress was to require elimination of all possible risk arising from a renovation. Nor does TSCA explicitly require EPA to eliminate all possible risk from lead, nor would it be feasible to do so since lead is a component of the earth. Rather, it directs EPA to regulate renovation and remodeling activities that create lead-based paint hazards. Given that the trigger for regulating renovation and remodeling activities is the creation of lead-based paint hazards—which EPA has identified in a separate rulemaking pursuant to TSCA section 403—EPA believes taking safety into account in this context is best interpreted with reference to those promulgated hazard standards.

If taking safety into account required a more stringent standard, as suggested by some commenters, the potential would be created for a scheme under which any renovation activities found not to create hazards are not regulated at all, whereas renovation activities found to create hazards trigger requirements designed to leave the renovation site cleaner than the unregulated renovations. EPA's interpretation is supported by the broad Congressional intent that the section 403 hazard standards apply for the purposes of subchapter IV of TSCA. It is also consistent with EPA's approach in its abatement regulations, which require post-abatement cleaning to dust-lead clearance levels that are numerically equal to the TSCA section 403 hazard standards levels.

It would be anomalous to impose a more stringent safety standard in the renovation context than in the abatement context, where the express purpose of the regulated activities is to abate lead-based paint hazards. Therefore, in taking into account safety, this final rule regulates renovation and remodeling activities relative to the TSCA section 403 hazard standard, with the purpose of minimizing exposure to such hazards created during renovation and remodeling activities.

Additionally, EPA has interpreted practicality in implementation to be an element of the statutory directive to take into account effectiveness and reliability. In particular, EPA believes that given the highly variable nature of the regulated community, the work practices required by this rule should be simple to understand and easy to use.

Those Affected by the New Rule

EPA has an explanation for who is affected by the new rule. The agency is very aware that this regulation will apply to a whole range of individuals from day laborers to property maintenance staff to master craftsmen performing a whole range of activities from simple drywall repair or window replacement to complete kitchen and bath renovations, to building additions and everything in-between. Work practices that are easy and practical to use are more likely to be followed by all of the persons who perform renovations, and, therefore, more likely to be reliable and effective in minimizing exposure to lead-based paint hazards created by renovation activities.

One of the biggest challenges facing EPA in revising the TSCA section 402(a) Lead-Based Paint Activities Regulations is how to effectively bridge the differences between abatement and renovation and remodeling while acknowledging that many of the dust generating activities are the same. Abatements are generally performed in three circumstances. First, an abatement may be performed in the residence of a child who has been found to have an elevated blood-lead level. Second, abatements are performed in housing receiving HUD financial assistance when required by HUD's Lead-Safe Housing Rule. Third, state and local laws and regulations may require abatements in certain situations associated with rental housing.

Typically, when an abatement is performed, the housing is either unoccupied or the occupants are temporarily relocated to lead-safe housing until the abatement has been demonstrated to have been properly completed through dust clearance testing. Carpet in the housing is usually removed as part of the abatement because it is difficult to demonstrate that it is free of lead-based paint hazards. Uncarpeted floors that have not been replaced during the abatement may need to be refinished or sealed in order to achieve clearance. Abatements have only one purpose: to permanently eliminate lead-based paint and lead-based paint hazards.

On the other hand, renovations are performed for myriad reasons, most of them having nothing to do with lead-based paint. Renovations involve activities designed to update, maintain, or modify all or part of a building. Renovations may be performed while the property is occupied or unoccupied. If the renovation is performed while the property is occupied, the occupants do not typically relocate pending the completion of the project.

Further, performing abatement is a highly specialized skill that workers and supervisors must learn in training courses accredited by EPA or authorized states, territories, or tribes. In contrast, EPA is not interested in teaching persons how to be painters, plumbers, or carpenters. Rather, EPA's objective is to ensure that persons who already know how to perform renovations perform their typical work in a lead-safe manner.

Nevertheless, as pointed out by some commenters, abatement and renovation have some things in common. For example, as noted by one commenter, window replacement may be performed as part of an abatement to remove the lead-based paint and lead-based paint hazards on the existing window, or it may be performed as part of a renovation designed to improve the energy efficiency of the building. In many cases, the window replacement as abatement and the window replacement as renovation will generate the same amount of leaded dust.

Another consideration is that while renovation activities undoubtedly create lead-based paint hazards, without results from dust wipe samples collected immediately before the renovation commences, there is no way to tell what portion of the lead dust remaining on the surface was contributed by the renovation. In addition, as a practical matter, once dust-lead hazards commingle with preexisting hazards, there is no functional way to distinguish between those created by the renovation activity and any preexisting dust-lead hazards. However, the Dust Study shows that the combination of training, containment, cleaning, and cleaning verification required by this rule is effective at reducing dust lead levels below the dust-lead hazard standard. While the requirements of this rule will, in some cases, have the ancillary benefit of removing some preexisting dust-lead hazards, these requirements are designed to effectively clean up the lead-based paint hazards created during renovation activities without changing the scope of the renovation activity itself. The intent of this final rule is not to require cleanup of preexisting contamination.

For example, the rule does not require cleaning of dust or any other possible lead sources in portions of target housing or child-occupied facilities beyond the location in and around the work area. Nor does this rule require the replacement of carpets in the area of the renovation or the refinishing or sealing of uncarpeted floors. The approach in this final rule is designed to address the lead-based paint hazards created during the renovation while not requiring renovators to remediate or eliminate hazards that are beyond the scope of the work they were hired to do.

Keeping Costs Low

EPA has made a concerted effort to keep the costs and burdens associated with this rule as low as possible, while still providing adequate protection against lead-based paint hazards created by renovation activities. Indeed, as part of this rulemaking EPA has, as directed by TSCA section 2(c), considered the environmental, economic, and social impact of this rule. Nonetheless, many commenters expressed concerns over the potential unintended consequences of this rulemaking. These commenters argued that a too-burdensome rule will result in more renovations by non-compliant renovators and more do-it-yourself renovations, both of which are likely to be more hazardous than renovations by certified professional renovation firms using certified renovators who follow the work practice requirements of the rule. These commenters were also concerned about deferred property maintenance which can be hazardous for many reasons, including lead-based paint issues. For example, one commenter pointed out that a renovation project that replaces old lead-based paint covered windows with new ones that have no lead paint may, as a by-product, reduce lead hazards, and the rule should not work to discourage this activity.

On the other hand, one commenter argued that increased do-it-yourself activity is an unlikely by-product of this rule because consumers are not only opting to hire or not hire contractors based on factors such as cost, convenience, and perceived quality, but, even more importantly, their own proclivity toward performing renovation work. According to the commenter, the fact that the work practices required by this rule may result in slight cost increases is unlikely to motivate homeowners to perform their own renovations. This commenter also felt that the sooner that protective approaches become the accepted standard of care for renovation work by contractors receiving compensation, the sooner do-it-yourselfers and the do-it-yourself literature and training supports will adopt the same protective approaches.

It is difficult to determine with any amount of certainty whether this final rule will have unintended consequences. However, EPA agrees that it is important to minimize disincentives for using certified renovation firms who follow the work practices required by this rule. EPA also agrees that practicality is an important consideration. Given the relatively low estimated overall average per-job cost of this final rule, which is $35, and the relatively easy-to-use work practices required by this final rule, EPA does not expect the incremental costs associated with this rule to be a determinative factor for consumers. However, that relatively low cost has resulted in part from EPA's efforts to contain the costs of this rule to avoid creating disincentives to using certified renovation firms, and EPA has viewed the comments received with those considerations in mind.

With respect to the comment regarding the standard of care for do-it-yourselfers, EPA also plans to conduct an outreach and education campaign aimed at encouraging homeowners and other building owners to follow work practices while performing renovations or to hire a certified renovation firm to do so.

Do you think we are finally done with the Lead Rule? No, we are not even close to being done. There is a lot to learn and we will continue our trek in the next chapter.

Structure of EPA Lead Renovation, Repair, and Painting Program

6

You have just read about how the EPA Lead Renovation, Repair, and Painting Program came to be. That is the tip of the iceberg. This ruling from EPA is very comprehensive and has generated a lot of talk in the construction trades. Like it or not, the rules must be followed when working with lead-based materials. Here we examine some of the more specific requirements of the rule.

▶ PRE-RENOVATION EDUCATION RULE

EPA developed a new renovation-specific lead hazard information pamphlet intended for use in fulfilling the requirements of the Pre-Renovation Education Rule, 40 CFR, Part 745, Subpart E. This final rule requires firms performing renovations for compensation in target housing and child-occupied facilities (COF) to distribute this new pamphlet before beginning renovations to the owners and occupants of target housing, owners of public or commercial buildings that contain a child-occupied facility, and the proprietor of the COF, if different, and to provide general information on the renovation and the pamphlet to, or make it available to, parents or guardians of children under age 6 using the child-occupied facility.

This can be accomplished by mailing or hand-delivering the general information on the renovation and the pamphlet to the parents and guardians or by posting informational signs containing general information on the renovation in areas where the signs can be seen by the parents or guardians of the children frequenting the COF. The signs must be accompanied by a posted copy of the pamphlet or information on how interested parents or guardians can review a copy of the pamphlet or obtain a copy from the renovation firm at no cost to the parents or guardians. For renovations in the common areas of multi-unit target housing, similar notification options are available to firms. They must provide tenants with general information regarding the nature of the renovation by mail, by hand-delivery, or by posting signs, and must also make this new pamphlet available on request. Firms must maintain documentation of compliance with these requirements.

Training, Accreditation, and Certification

This final rule contains training requirements leading to certification for "renovators"—individuals who perform and direct renovation activities—and "dust sampling technicians"—individuals who perform dust sampling not in connection with an abatement. Requirements for each course of study are described in detail and a hands-on component is required. Training providers who wish to provide training to renovators and dust sampling technicians for federal certification purposes must apply for and receive accreditation from EPA following the same procedures that training providers who offer lead-based paint activities training now use to become accredited by EPA.

Providers of renovation training must follow the same requirements for program operation as training providers who offer lead-based paint activities training. For example, renovation training programs must have adequate facilities and equipment for delivering the training, a training manager with experience or education in a construction or environmental field, and a principal instructor with experience or education in a related field and education or experience in teaching adults. To become accredited to provide training for renovators and dust sampling technicians, a provider must submit an application for accreditation to EPA. The application must include the following items:

- The course materials and syllabus, or a statement that EPA model materials or materials approved by an authorized state or tribe will be used
- A description of the facilities and equipment that will be used
- A copy of the test blueprint for each course
- A description of the activities and procedures that will be used during the hands-on skills portion of each course
- A copy of the quality control plan
- The correct amount of fees

Training programs that submit a complete application and meet the requirements for faculty, facilities, equipment, and course and test content will be accredited for 4 years. To maintain accreditation, the training program must submit an application and the correct amount of fees every 4 years. EPA is not establishing the required fees in this rulemaking; it intends to publish a proposed fee schedule for public comment shortly.

Accredited renovation training programs must also comply with the existing notification and recordkeeping requirements for lead-based paint activities training programs at 40 CFR 745.225(c)(13) and 40 CFR 745.225(i), respectively, by notifying EPA before and after providing renovation training and by maintaining records of course materials, course test blueprints, information on how hands-on training is delivered, and the results of the students' skills assessments and course tests.

Each renovation project covered by this final rule must be performed and/or directed by an individual who has become a certified renovator by successfully completing renovator training from an accredited training provider. The certified renovator is responsible for ensuring compliance with the work practice standards of this final regulation. The certified renovator must perform or direct certain critical tasks during the renovation, such as posting warning signs, establishing containment of the work area, and cleaning the work area after the renovation.

These and other renovation activities may be performed by workers who have been provided on-the-job training in these activities by a certified renovator. However, the certified renovator must be physically present at the worksite while signs are being posted, containment is being established, and the work area is being cleaned after the renovation to ensure that these tasks are performed correctly. Although the certified renovator is not required to be onsite at all times while the renovation project is ongoing, a certified renovator must nonetheless regularly direct the work being performed by other workers to ensure that the work practices are being followed.

When a certified renovator is not physically present at the worksite, the workers must be able to contact the renovator immediately by telephone or other mechanism. In addition, the certified renovator must perform the post-renovation cleaning verification. This task may not be delegated to workers with on-the-job training. To maintain certification, a renovator must successfully complete an accredited renovator refresher training course every 5 years.

Renovations must be performed by certified firms. The certification requirements for renovation firms are identical to the certification requirements for firms that perform lead-based paint activities, except that renovation firm certification lasts for 5 years instead of 3 years. A firm that wishes to become certified to perform renovations must submit an application, along with the correct amount of fees, attesting that it will assign a certified renovator to each renovation that it performs, that it will use only certified or properly trained individuals to perform renovations, and that it will follow the work practice standards and recordkeeping requirements in this regulation.

EPA will certify any firm that meets these requirements unless it determines that the environmental compliance history of the firm, its principals, or its key employees demonstrates an unwillingness or inability to maintain compliance with environmental statutes or regulations. To maintain certification, the firm must submit an application and the correct amount of fees every 5 years. As noted above, EPA will establish the required fees in a subsequent rulemaking.

Work Practice Standards

This final rule contains a number of work practice requirements that must be followed for every covered renovation in target housing and COFs. These requirements pertain to warning signs and work area containment, the restriction or prohibition of certain practices (e.g., high heat gun, torch, power sanding, power planing), waste handling, cleaning, and post-renovation cleaning verification. The firm must ensure compliance with these work practices. Although the certified renovator is not required to be onsite at all times while the renovation project is ongoing, a certified renovator must nonetheless regularly direct the work being performed by other workers to ensure that the work practices are being followed. When a certified renovator is not physically present at the worksite, the workers must be able to contact the renovator immediately by telephone or other mechanism.

Warning Signs and Work Area Containment

Before beginning a covered renovation, the certified renovator or a worker under the direction of the certified renovator must post signs outside the area to be renovated warning occupants and others not involved in the renovation to remain clear of the area. In addition, the certified renovator or a worker under the direction of the certified renovator must also contain the work area so that dust or debris does not leave the area while the work is being performed. At a minimum, containment for interior projects must include:

- Removing or covering all objects in the work area with plastic or other impermeable material
- Closing and covering all forced air HVAC ducts in the work area with plastic or other impermeable material
- Closing all windows in the work area
- Closing and sealing all doors in the work area with plastic or other impermeable material
- Covering the floor surface, including installed carpet, with taped-down plastic sheeting or other impermeable material in the work area 6 feet beyond the perimeter of surfaces undergoing renovation or a sufficient distance to contain the dust, whichever is greater

Doors within the work area that will be used while the job is being performed must be covered with plastic sheeting or other impermeable material in a manner that allows workers to pass through while confining dust and debris to the work area. In addition, all personnel, tools, and other items, including the exterior of containers of waste, must be free of dust and debris when leaving the work area. There are several ways of accomplishing this. For example, tacky mats may be put down immediately adjacent to the plastic sheeting covering the work area floor to remove dust and debris from the bottom of the workers' shoes as they leave the work area, workers may remove their shoe covers (booties) as they leave the work area, and clothing and materials may be wet-wiped and/or HEPA-vacuumed before they are removed from the work area.

At a minimum, containment for exterior projects must include covering the ground with plastic sheeting or other disposable impermeable material extending 10 feet beyond the perimeter of surfaces undergoing renovation or a sufficient distance to collect falling paint debris, whichever is greater, unless the property line prevents 10 feet of such ground covering. Closing all doors and windows within 20 feet of the outside of the work area on the same floor as the renovation and closing all doors and windows on the floors below that area is required.

In certain situations, such as where other buildings are in close proximity to the work area, when conditions are windy, or where the work area abuts a property line, the certified renovator or a worker under the direction of the certified renovator performing the renovation may have to take extra precautions to prevent dust and debris from leaving the work area as required by the regulation. This may include erecting a system of vertical containment designed to prevent dust and debris from migrating to adjacent property or contaminating the ground, other buildings, or any object beyond the work area. In addition, doors within the work area that will be used while the job is being performed must be covered with plastic sheeting or other impermeable material in a manner that allows workers to pass through while confining dust and debris to the work area.

Waste Management

The certified renovator or a worker trained and directed by a certified renovator must, at the conclusion of each work day, store any collected lead-based paint waste from renovation activities under containment, in an enclosure, or behind a barrier that prevents release of dust and debris and prevents access to the waste. In addition, the certified renovator or a worker under the direction of the certified renovator transporting lead-based paint waste from a worksite must contain the waste to prevent identifiable releases.

With regard to the lead-based paint waste generated by renovations in housing units, Unit IV.D.2. of the preamble to the 2006 Proposal describes how a clarification of the hazardous waste exclusion in 40 CFR 261.4(b)(1) means that residential lead-based paint waste may be disposed of in municipal solid waste landfill units, as long as the waste is generated during abatement or renovation and remodeling activities in households. Also discussed in the preamble to the 2006 Proposal is a subsequent amendment to the waste regulations promulgated under the Resource Conservation and Recovery Act (RCRA) that allows construction and demolition (C&D) landfills to accept residential lead-based paint waste.

Cleaning

This final rule contains a number of specific cleaning steps that the certified renovator or a worker under the direction of the certified renovator must follow after performing a covered renovation. On completion of renovation activities, all paint chips and debris must be picked up. Protective sheeting must be misted and folded dirty side inward. Sheeting used to isolate the work area from other areas must remain in place until after the cleaning and removal of other sheeting; this sheeting must be misted and removed last. Removed sheeting must either be folded and taped shut to seal or sealed in heavy-duty bags and disposed of as waste.

After the sheeting has been removed from the work area, the entire area must be cleaned, including the adjacent surfaces that are within 2 feet of the work area. The walls, starting from the ceiling and working down to the floor, must be vacuumed with a HEPA vacuum or wiped with a damp cloth. This final rule requires that all remaining surfaces and objects in the work area, including floors, furniture, and fixtures, be thoroughly vacuumed with a HEPA-equipped vacuum. When cleaning carpets, the HEPA vacuum must be equipped with a beater bar to aid in dislodging and collecting deep dust and lead from carpets. Where feasible, floor surfaces underneath area rugs must also be thoroughly vacuumed with a HEPA vacuum.

After vacuuming, all surfaces and objects in the work area, except for walls and carpeted or upholstered surfaces, must be wiped with a damp cloth. Uncarpeted floors must be thoroughly mopped using a 2-bucket mopping method that keeps the wash water separate from the rinse water, or using a wet mopping system with disposable absorbent cleaning pads and a built-in mechanism for distributing or spraying cleaning solution from a reservoir onto a floor.

For cleaning following an exterior renovation, this final rule requires all paint chips and debris to be picked up. Protective sheeting must be misted and folded dirty side inward. Removed sheeting must be either folded and taped shut to seal or sealed in heavy-duty bags and disposed of as waste.

Post-Renovation Cleaning Verification

This final rule requires a certified renovator to perform a visual inspection of the work area after the cleaning steps outlined in the previous subsection. This visual inspection is for the purpose of determining whether dust, debris, or other residue is present in the work area. If dust, debris, or other residue remains in the work area, the dust, debris, or other residue must be removed by recleaning and another visual inspection must be performed.

When an exterior work area passes the visual inspection, the renovation has been properly completed and the warning signs may be removed. When an interior work area passes the visual inspection, an additional cleaning verification step is required. A certified renovator assigned to the renovation project must use disposable cleaning cloths to wipe the windowsills, countertops, and uncarpeted floors in the work area. These cloths must then be compared to a cleaning verification card. For each cloth that matches or is lighter than the cleaning verification card, the corresponding windowsill, countertop, or floor area is considered to have passed the post-renovation cleaning verification. In contrast to the 2006 Proposal, this final rule limits this requirement to two wet cloths and one dry cloth. After the first dry cloth, that surface will be considered to have passed post-renovation cleaning verification. When all windowsills, countertops, and floor areas in the work area have passed post-renovation cleaning verification, the warning signs may be removed.

In contrast to the 2006 Proposal, this final rule does not allow dust clearance sampling in lieu of post-renovation cleaning verification, except in cases where the contract between the renovation firm and the property owner or another federal, state, territorial, tribal, or local regulation requires dust clearance sampling by a certified sampling professional and requires the renovation firm to clean the work area until it passes clearance.

▶ STATE, TERRITORIAL, AND TRIBAL PROGRAMS

This final rule also contains provisions for interested states, territories, and tribes to apply for and receive authorization to administer their own renovation, repair, and painting programs in lieu of the proposed regulation. States, territories, and tribes may choose to administer and enforce just the existing requirements of Subpart E, the pre-renovation education elements, the training, certification, accreditation, work practice, and record-keeping requirements of this final rule, or both. EPA will use the same process used for lead-based paint activities programs, along with proposed specific renovation program elements, to authorize state, territorial, and tribal programs.

States, territories, and tribes seeking authority to administer and enforce renovation programs must obtain public input and then submit an application to EPA. Applications must contain a number of items, including a description of the state, territorial, or tribal program, copies of all applicable statutes, regulations, and standards, and a certification by the State Attorney General, Tribal Counsel, or an equivalent official, that the applicable legislation and regulations provide adequate legal authority to administer and enforce the program. The program description must demonstrate that the state, territorial, or tribal program is at least as protective as the federal program and that it provides for adequate enforcement.

To be eligible for authorization to administer and to enforce renovation programs, state, territorial, and tribal renovation programs must contain certain minimum elements that are very similar to the minimum elements required for lead-based paint activities programs. To be authorized, state, territorial, or tribal programs must have procedures and requirements for the accreditation of training programs, the training of renovators, and the certification of renovators or renovation firms.

At a minimum, the program requirements must include accredited training for renovators and procedures and require-ments for recertification. State, territorial, and tribal programs applying for authorization are also required to include work practice standards for renovations that ensure that renovations are conducted only by certified renovators or renovation firms and that renovations are conducted using work practices at least as protective as those of the Federal program.

Renovations Affecting Only Components Free of Regulated Lead-Based Paint

In keeping with the 2006 Proposal and the 2007 Supplemental Proposal, this final rule exempts renovations that affect only components that a certified inspector or risk assessor has deter-mined are free of paint or other surface coatings that contain lead equal to or in excess of $1.0\,mg/cm^2$ or 0.5% by weight. These standards are from the definition of lead-based paint in Title X and in EPA's implementing regulations. Nearly all of the commenters that expressed an opinion on this topic favored this exception. The determination that any particular component is free of lead-based paint may be made as part of a lead-based paint inspection of an entire housing unit or building, or on a component-by-component basis.

Some commenters expressed confusion over the mechanics of this exception. The certified inspector or risk assessor determines whether components contain lead-based paint, while the renovation firm is responsible for determining which components will be affected by the renovation. A renovation firm may rely on the report of a past inspection or risk assessment that addresses the components that will be disturbed by the renovation.

Determination by a Certified Renovator Using EPA-Recognized Test Kits

This final rule exempts renovations that affect only components that a certified renovator, using a test kit recognized by EPA, determines are free of lead-based paint. EPA has deleted the regulatory thresholds for lead-based paint from this definition because they unnecessarily complicate the exception. A certified renovator is a person who has taken an accredited course in work practices. This training will include how to properly use the EPA-approved test kits. This final rule also establishes the process EPA will use to recognize test kits.

As discussed in the preamble to the 2006 Proposal, research on the use of currently available kits for testing lead in paint has been published by the National Institute of Standards and Technology (NIST). The research indicates that there are test kits on the market that, when used by a trained professional, can reliably determine that regulated lead-based paint is not present by virtue of a negative result. Based on this research, EPA proposed to initially recognize test kits that have, for paint containing lead at or above the regulated level, $1.0\,mg/cm^2$ or 0.5% by weight, a demonstrated probability (with 95% confidence) of a negative response less than or equal to 5% of the time.

Some commenters, representing a variety of interests, supported an exception for renovations affecting components that have been found to be free of regulated lead-based paint by use of a test kit. One commenter cited the need for faster and cheaper methods of accurately checking for lead and expressed the opinion that this approach will expand access to lead screening in homes. Several comments were generally supportive, with some reservations about kit reliability.

However, most commenters did not favor the use of test kits. The most commonly cited reason for not supporting this approach was the potential conflict-of-interest present in having the certified renovator be the one to determine whether or not he or she must use the work practices required by the rule. EPA addressed potential conflicts of interest in its lead-based paint program in the preamble to the final Lead-Based Paint Activities Regulations. That discussion outlined two reasons for not requiring that inspections or risk assessments, abatements, and post-abatement clearance testing all be performed by different entities.

The first was the cost savings and convenience of being able to hire just one firm to perform all necessary lead-based paint activities. The second was the potential regional scarcity of firms to perform the work. These considerations may also be applicable to the renovation sector, given the premium on maintaining a rule that is simple and streamlined and does not unduly prolong the timeframes for completing renovations. Moreover, it is not unusual in regulatory programs to allow regulated entities to make determinations affecting regulatory applicability and compliance. EPA has decided to take an approach that is consistent with the approach taken in the 402(a) Lead-Based Paint Activities Regulations and not require third-party testing.

Another commonly cited reason for not supporting the use of test kits by certified renovators was the lack of any sampling protocol in the regulation. A related concern was that the training in sampling techniques and protocols in the lead-based paint inspector course could not be shortened to fit within the 8-hour renovator course and still retain all of the necessary information. EPA wishes to make it clear that the 8-hour renovator course will not train renovators in how to select components for sampling because the certified renovator must use a test kit on each component affected by the renovation. The only exception to this is when the components make up an integrated whole, such as the individual stair treads and risers in a staircase.

In this situation, the renovator need test only one such individual component (e.g., a single stair tread) unless it is obvious to the renovator that the individual components have been repainted or refinished separately. As such, a complicated sampling protocol is not necessary. EPA plans to modify the EPA/HUD Lead-Safe Work Practices course to include training on how to use a test kit. To ensure that the applicability of the exception is clear, EPA has also modified 40 CFR 745.82(a)(2) to specifically state that the certified renovator must test each of the components that will be affected by the renovation.

▶ PHASED IMPLEMENTATION AND IMPROVED TEST KITS

Under the proposals, the regulatory requirements would have taken effect in two major stages, based on the age of the building being renovated. The first stage would have applied to renovations in target housing and COFs built before 1960. Requirements for renovations in target housing and child-occupied facilities built between 1960 and 1978 would have taken effect 1 year later. The primary reason for this phased implementation was to allow time for the development of improved test kits.

According to the National Survey of Lead and Allergens in Housing, 24% of the housing constructed between 1960 and 1978 contains lead-based paint. In contrast, 69% of the housing constructed between 1940 and 1959, and 87% of the housing constructed before 1940 contains lead-based paint. The results of this survey indicate that there is a much greater likelihood of disturbing lead-based paint during a renovation that occurs in a home built before 1960 than in a home built after that date.

The NIST research on existing test kits shows that existing test kits cannot reliably determine that lead is present in paint only above the statutory levels because the kits are sensitive to lead at levels below the federal standards that define lead-based paint, and therefore are prone to a large number of false positive results (i.e., a positive result when regulated lead-based paint is, in fact, not present). The NIST research found that such false positive rates range from 42 to 78%. This means that the currently available kits are not an effective means of identifying the 76% of homes built between 1960 and 1978 that do not contain regulated lead-based paint.

Research conducted by EPA subsequent to the publication of the 2006 Proposal confirms that the sensitivity of test kits could be adjusted for paint testing so that the results from the kits reliably correspond to one of the two federal standards for lead-based paint, $1.0\,mg/cm^2$ and 0.5% by weight. EPA's research and initial contacts with potential kit manufacturers also indicate that this can be accomplished in the near future. As stated in the preamble to the 2006 Proposal, EPA's goal is to foster the development of a kit that can reliably be used by a person with minimal training, is inexpensive, provides results within an hour, and is demonstrated to have a false positive rate of no more than 10% and a false negative rate at $1.0\,mg/cm^2$ or 0.5% by weight of less than 5%. EPA planned to have improved test kits meeting EPA's benchmarks commercially available by September 2010.

With this in mind, EPA felt that a staged approach would initially address the renovations that present the greatest risks to children under age 6 (i.e., the renovations that are most likely to disturb lead-based paint), while allowing additional time to ensure that the improved test kits are commercially available before phasing in the applicability of the rule to newer target housing and child-occupied facilities. However, EPA was concerned about delaying implementation for post-1960 target housing and COFs that are occupied or used by children under age 6 with increased blood-lead levels.

To reduce the possibility that an unregulated renovation activity would contribute to continuing exposures for these children, the 2006 Proposal would have required renovation firms, during the first year that the training, certification, work practice, and recordkeeping requirements are in effect, to provide owners and occupants of target housing built between 1960 and 1978 and child-occupied facilities built between 1960 and 1978 the opportunity to inform the firm that the building to be renovated is the residence of, or is a child-occupied facility frequented by, a child under age 6 with a blood-lead level that equals or exceeds the CDC level of concern, or a lower state or local government level of concern. If the owner or occupant informs the renovation firm that a child under age 6 with an increased blood-lead level lives in or frequents the building to be renovated, the renovation firm must comply with all of the training, certification, work practice, and recordkeeping requirements of this regulation.

Some commenters agreed that a staged approach was probably necessary, given the number of renovations that would be covered by the rule, and that a focus on buildings built before

1960 was appropriate. However, most commenters objected to the phased implementation. Some were concerned about the potential exposures to children in buildings built between 1960 and 1978 during the first stage of the rule.

Another major concern expressed by commenters was that the phased implementation would unnecessarily complicate the rule, especially with the provision relating to children under age 6 with increased blood-lead levels. These commenters felt that, because there already are accurate methods for determining whether a building contains lead-based paint, and because renovation firms ought to get into the habit of working in a lead-safe manner whenever they are working on a building built before 1978, the utility of the delay does not outweigh the likely confusion in the regulated community. Commenters also expressed reservations about providing sensitive medical information to contractors, in the case of children under age 6 with increased blood-lead levels.

After reviewing the comments and weighing all of the factors, including EPA's expectation that the improved test kits are now commercially available, EPA decided not to include a phased implementation in this rulemaking. Therefore, this regulation will take effect at the same time for target housing and child-occupied facilities regardless of whether they were built before or after 1960. Nonetheless, if the improved test kits have not been made commercially available, EPA will initiate a rulemaking to extend the effective date of this final rule for 1 year with respect to owner-occupied target housing built after 1960.

Test Kit Recognition Process

In the 2006 Proposal, EPA described proposed criteria for test kit recognition. Specifically, for paint containing lead at or above the regulated level, $1.0 \, mg/cm^2$ or 0.5% by weight, EPA stated its intention to only recognize kits that have a demonstrated probability (with 95% confidence) of a negative response less than or equal to 5% of the time. In addition, as soon as the improved test kits are generally available, EPA will recognize only those test kits that have a demonstrated probability (with 95% confidence) of a false positive response of no more than 10% to lead in paint at levels below the regulated level.

EPA stated its belief that limiting recognition to kits that demonstrate relatively low rates of false positives would benefit the consumer by reducing the number of times that the training and work practice requirements of this regulation are followed in the absence of regulated lead-based paint. EPA also proposed to require that these performance parameters be validated by a laboratory independent of the kit manufacturer, using ASTM International's E1828, Standard Practice for Evaluating the Performance Characteristics of Qualitative Chemical Spot Test Kits for Lead in Paint or an equivalent validation method.

In addition, the instructions for use of any particular kit would have to conform to the results of the validation, and the certified renovator would have to follow the manufacturer's instructions when using the kit. EPA requested comment on whether these standards are reasonably achievable and sufficiently protective. EPA also solicited input on how to conduct the kit recognition process.

Some commenters expressed reservations about the proposed performance criteria, contending that a false negative rate of 5% is too high to be protective. However, a 5% false negative rate (with 95% confidence) is similar to the performance requirements for other lead-based paint testing methods, such as laboratory analysis used for lead-based paint inspections, and is considered to be the statistical equivalent of zero. Therefore, this final rule retains the proposed false-negative criteria for test kit recognition, that is, for paint containing lead at or above the regulated level, $1.0\,mg/cm^2$, or 0.5% by weight, kits will be recognized only if they have a demonstrated probability (with 95% confidence) of a negative response less than or equal to 5% of the time.

Because no comments were received on the proposed false-positive criteria of 10% for the improved test kits, this final rule also retains the proposed false-positive criteria for the improved kits, that is, after the improved kits are available, the only test kits that will be recognized are those that have a demonstrated probability (with 95% confidence) of a false positive response of no more than 10% to lead in paint at levels below the regulated level.

EPA did not receive any comments or suggestions on the test kit recognition process itself. With respect to existing test kits, EPA has determined that the NIST research is the equivalent of an independent laboratory validation of test kit performance.

The NIST research found that three kits met the false-negative criteria established in this final rule. For the purposes of this regulation, EPA will therefore recognize these test kits, provided that they still use the same formulation that was evaluated by NIST.

With respect to the improved test kits, EPA has determined that Environmental Technology Verification Program (ETV) is a suitable vehicle for obtaining independent laboratory validation of test kit performance. EPA intends to use ETV or an equivalent testing program approved by EPA for the test kit recognition process. The goal of the ETV Program is to provide independent, objective, and credible performance data for commercial-ready environmental technologies.

The ETV process promotes these technologies' implementation for the benefit of purchasers, permitters, vendors, and the public. If ETV is used, EPA would utilize the Environmental and Sustainable Technology Evaluations (ESTE) element of the ETV program because the development of the test kits is in support of this final rule, and the ESTE element was created in 2005 to address the agency's priorities such as rule making. More information on this program is available on EPA's website at *www.epa.gov/etv/index.html*.

In the 2006 Proposal, EPA noted that it would look to ASTM International's E1828, Standard Practice for Evaluating the Performance Characteristics of Qualitative Chemical Spot Test Kits for Lead in Paint, or equivalent for a validation method for test kits. With the input of stakeholders, EPA is adapting this ASTM Standard for use in the laboratory validation program. The testing protocol will consist of an evaluation of the performance of the test kits, using the manufacturer's instructions, on various substrates, such as wood, steel, drywall, and plaster, with various lead compounds, such as lead carbonate and lead chromate, at various lead concentrations above and below regulatory threshold for lead-based paint.

To be consistent with the performance criteria of the National Lead Laboratory Accreditation Program, the testing protocol will not involve testing the performance of the kits on paint that contains between 0.8 milligrams of lead per square centimeter and 1.2 milligrams of lead per square centimeter. After a test kit has gone through the ETV or other EPA approved testing process, EPA will review the test report to determine whether the kit has been demonstrated to achieve the criteria set forth in the rule. EPA anticipated that evaluation of the improved test kits under the recognition program would begin by August 2009.

In addition, EPA intends to allow other existing test kit manufacturers the opportunity to demonstrate that their kits meet the false negative criteria described in 40 CFR 745.88(c)(1) by going through the ETV process. Any recognition granted to test kits based only on the false negative criteria will expire when EPA publicizes its recognition of the first improved test kit that meets both the false negative and false positive criteria of 40 CFR 745.88(c).

Beginning on September 1, 2008, EPA's ETV program was to accept applications for testing from test kit manufacturers. Applications must be submitted, along with a sufficient number of kits and the instructions for using the kits, to EPA. The test kit manufacturer should first visit the following website for information on where to apply: *www.epa.gov/etv/howtoapply.html*.

▶ MINOR REPAIR AND MAINTENANCE

EPA proposed to incorporate into this regulation the minor maintenance exception for the Pre-Renovation Education Rule. The proposed minor maintenance exception would have applied to projects that disturb 2 ft^2 or less of painted surface per component. The preamble to the 2006 Proposal discusses the history of this exception and requested comment on potential changes.

In particular, EPA noted that HUD's Lead Safe Housing Rule, at 20 CFR 35.1350(d), includes a *de minimis* exception for projects that disturb 2 ft^2 or less of painted surface per room for interior projects, 20 ft^2 or less of painted exterior surfaces, and 10% or less of the total surface area on an interior or exterior type of component with a small surface area. If less than this amount of painted surface is disturbed, HUD's lead-safe work practice requirements do not apply. EPA's Lead-Based Paint Activities Regulation incorporates this as an exception for small projects at 40 CFR 745.65(d). EPA requested comment on whether the minor maintenance exception in this regulation should be consistent with other EPA regulations and the HUD Lead Safe Housing Rule. This provision describes the applicability of the Pre-Renovation Education Rule as well as this final rule.

Most commenters expressed support for consistency in the various lead-based paint regulations administered by EPA and HUD. They noted that a consistent exception for small projects or minor maintenance would be easier for the regulated community to apply. Many of these commenters recommended $2\,ft^2$ for interior projects and $20\,ft^2$ on exterior surfaces. While some commenters supported a "per component" exception, several commenters specifically noted that the "per component" aspect of the existing Pre-Renovation Education Rule exception was problematic in that it could result in the disturbance of large areas of painted surfaces in a single room. Other commenters recommended that the threshold area for the exception be made smaller or the exception abolished.

These commenters noted that even very small projects have the potential to create lead-based paint hazards and that, rather than worrying about the applicability of the exception, renovation firms should just get into the habit of performing every project in a lead-safe manner. Other commenters suggested that EPA consider a larger threshold area for the exception, or an exception based on other factors, such as time spent performing an activity. EPA recognizes that, depending on the methods used to disturb lead-based paint, very small disturbances can release a great deal of lead. EPA also understands the practicality of a minor maintenance exception.

In weighing these competing considerations, EPA has decided to incorporate in this final rule a minor maintenance exception for projects that disturb $6\,ft^2$ or less of painted surface per room for interiors and $20\,ft^2$ or less of painted surface on exteriors. This addresses the concerns of those commenters who supported a per-component exception while still limiting the overall amount of paint that can be disturbed in a single room during a single project. As in the 2006 Proposal, this exception is not available for window replacement projects. In contrast to the proposal, this exception is only available for projects that do not use any of the work practices prohibited or restricted by 40 CFR 745.85(a)(3) and that do not involve demolition of painted surface areas.

The Environmental Protection Agency remains convinced that the distinction between renovation and minor maintenance activities is an important part of implementing this program. Congress directed EPA to address renovation and remodeling. In ordinary usage, minor maintenance activities that might disturb lead-based paint (e.g., removing a face plate for an electric switch to repair a loose connection, adding a new cable TV outlet, or removing a return air grill to service the HVAC system) are not normally considered home renovations. EPA believes that minor repair and maintenance activities that cover 6 ft^2 or less per room and 20 ft^2 or less for exteriors and that do not involve prohibited practices, demolition, or window replacement would not ordinarily be considered renovation or remodeling but would better be described as minor work on the home or COF.

EPA also believes that a typical minor repair and maintenance activity would not normally involve the use of high-dust generating machinery such as those prohibited or restricted by this rule. To make the distinction between renovations and minor repair and maintenance activities clear, EPA has added a definition of "minor repair and maintenance activities" to 40 CFR 745.83. This term is defined as follows:

Minor repair and maintenance activities are activities, including minor heating, ventilation, or air-conditioning work, electrical work, and plumbing, that disrupt 6 square feet or less of painted surface per room for interior activities or 20 square feet or less of painted surface for exterior activities where none of the work practices prohibited or restricted by § 745.85(a)(3) are used and where the work does not involve window replacement or demolition of painted surface areas. When removing painted components, or portions of painted components, the entire surface area removed is the amount of painted surface disturbed. Jobs, other than emergency renovations, performed in the same room within the same 30 days must be considered the same job for the purpose of determining whether the job is a minor repair and maintenance activity.

To accommodate this new definition of "minor repair and maintenance activities," the definition of "renovation" in §745.83 has also been changed to include the following sentence: "The term renovation does not include minor repair and maintenance activities." As a result of these two definitional changes, the reference to minor maintenance in 40 CFR 745.82(a)(1) is no longer necessary. Therefore, when engaged in minor repair and maintenance activities as defined in 40 CFR 745.83, renovation firms and renovators are not covered by this rule. EPA believes this approach—eliminating the per-component limitation in favor of an overall size cap, and prohibiting practices that EPA believes are inconsistent with minor maintenance work and that generate very high lead dust loadings—is a reasonable balance of the considerations identified by commenters and considered by EPA.

Several commenters expressed concerns about how the exception would be applied, and whether various activities would be covered by the rule or exempt under the minor maintenance exception. Window replacement was of interest to several commenters, who referred to EPA's previous guidance on window replacement under the Pre-Renovation Education Rule. That guidance states that window replacement, for various reasons, cannot qualify for the minor maintenance exception.

EPA knows of no reason why this interpretation should be changed. In fact, contrary to the assertions of some commenters, the Dust Study found that window replacement was one of the more hazardous jobs. The geometric mean of the lead content of floor dust samples taken in the work area after the window replacement projects was $3003 \, \mu g/ft^2$. In addition, EPA does not believe that window replacement is within the common understanding of the meaning of either minor repair or maintenance. EPA has specifically included language in the definition of "minor repair and maintenance activities" to make it clear that window replacements cannot qualify.

Two commenters contended that, when determining whether wall or ceiling cut-outs exceed the minor maintenance exception, the painted surface disturbed should be measured by multiplying the length of the cut by its width, as opposed to the total size of the cut-out. EPA disagrees with these commenters. For cut-outs, the calculation is made for the entire area of surface being disturbed, for example, the area of the cut-out, for the following reasons: the removed portion can flex or be broken during the removal process and the paint can flake off. The removed portion can fall on the floor and be trampled on, or the removed portion may not be removed as a single piece.

Calculating the amount of painted surface disturbed in the manner that the commenters suggested would also complicate the rule and be more difficult to convey during the renovator training course. In response to these comments, EPA has inserted clarifying language on this into the text of the definition of "minor repair and maintenance activities" at 40 CFR 745.83.

One commenter recommended that EPA prohibit splitting work, that is, conducting a single project as several minor maintenance activities in the same room in a short time (like a month) in order to avoid the regulatory requirements. EPA agrees with this commenter.

It has always been EPA's interpretation of the Pre-Renovation Education Rule that renovators could not artificially split up projects to avoid having to provide the pamphlet. In response to this comment, EPA has inserted clarifying language on this into the definition of "minor repair and maintenance activities" at 40 CFR 745.83. This definition states that jobs, other than emergency renovations, performed in the same room within the same 30 days must be considered the same job for the purpose of determining whether the job is a minor repair and maintenance activity.

▶ EMERGENCY PROJECTS

Both the 2006 Proposal and the 2007 Supplemental Proposal proposed to retain the emergency project exception in the Pre-Renovation Education Rule with one modification. EPA proposed to clarify that interim control projects performed on an expedited basis in response to an elevated blood-lead level finding in a resident child qualify for the emergency project exception from the Pre-Renovation Education Rule requirements. As discussed in the 2006 Proposal, EPA was concerned that local public health organizations may be delayed in responding to a lead-poisoned child if the owner of the building where the child resides is not available to acknowledge receipt of the lead hazard information pamphlet before an interim control project begins.

In addition, EPA recognized that some emergencies could make it difficult to comply with all of the training, certification, work practice, and recordkeeping requirements. For example, a broken water pipe may make it impossible to contain the work area before beginning to disturb painted surfaces to get to the pipe. The proposed emergency project exception would have required firms to comply with the work practice, training, certification, and recordkeeping requirements to the extent practicable.

EPA received a number of comments on this aspect of the 2006 Proposal. Several recognized the need for such an exception, but most of the commenters were concerned that the language of the proposal would make it possible for renovation firms to circumvent the training, certification, and work practice controls when performing interim controls in response to a child with an elevated blood-lead level. A number of these commenters, as well as several others, urged EPA to be more specific about which requirements could be bypassed in particular situations. EPA agrees with these commenters. It never was EPA's intention to allow firms performing interim controls in response to a poisoned child to use untrained workers or work in a manner not consistent with the work practices required by this rule.

EPA has therefore revised the exception to specifically state that interim controls performed in response to a child with an elevated blood-lead level are only exempt from the information distribution requirements, which is consistent with the current Pre-Renovation Education Rule. The EPA has also modified the exception to state that emergency renovations are only exempt to the extent necessary to respond to the emergency from the training, certification, sign posting, and containment requirements of this regulation. For example, most property management companies who do their own maintenance are likely to have at least one trained and certified renovator on staff to perform renovations, so these companies should be able to comply with the training and certification requirements on all renovations.

Likewise, firms performing emergency renovations should be able to follow the required cleaning procedures after emergency repairs have been made. As such, under the final rule, in all cases the cleaning specified by the regulation must be performed and it must be performed or directed by certified renovators. In addition, in all cases, the cleaning verification requirements of this regulation must be performed and they must be performed by a certified renovator. In response to one commenter who requested that EPA require firms to document their inability to comply with all of the regulatory provisions in emergencies, EPA has included such a requirement in 40 CFR 745.86(b)(7).

Finally, EPA has removed the word "operations" from the exception, in response to one commenter who suggested that the word is unnecessary and confusing. EPA agrees that the word "operations" is unnecessary in its description of emergency renovations. EPA intends to continue interpreting the term "emergency renovations" in the same way that it always has done, except that EPA has clarified that interim controls performed in response to a child with an elevated blood-lead level can be an emergency renovation.

▶ PRE-RENOVATION EDUCATION

The Pre-Renovation Education Rule, promulgated pursuant to TSCA section 406(b) and codified at 40 CFR, Part 745, Subpart E, requires renovators to provide owners and occupants of target housing with a lead hazard information pamphlet before beginning a renovation in the housing. The pamphlet currently used for this purpose, *Protect Your Family from Lead in Your Home*, was developed in accordance with TSCA section 406(a) and includes useful information on lead-based paint and lead-based paint hazards in general. This pamphlet is also used to provide lead hazard information to purchasers and renters of target housing under the Requirements for Disclosure of Information Concerning Lead-Based Paint in Housing "Lead Disclosure Rule."

New Renovation-Specific Pamphlet

EPA has developed a new lead hazard information pamphlet that addresses renovation-specific lead exposure concerns. The development of this pamphlet, including the public comments received on the format and content, is discussed in greater detail in a separate notice published elsewhere in the *Federal Register*. This renovation-specific pamphlet, entitled *Renovate Right: Important Lead Hazard Information for Families, Child Care Providers and Schools*, better informs families about the risks of exposure to lead-based paint hazards created during renovations and promotes the use of work practices and other health and safety measures during renovation activities.

This pamphlet gives information on lead-based paint hazards, lead testing, how to select a contractor, what precautions to take during the renovation, and proper cleanup activities, while still incorporating the information already included in the original *Protect Your Family from Lead in Your Home* and mandated by section 406(a) of TSCA.

In the 2006 Proposal, EPA proposed to require renovation firms to distribute the new renovation-specific pamphlet (then titled *Protect Your Family from Lead during Renovation, Repair, and Painting*) instead of the pamphlet currently used for this purpose (*Protect Your Family from Lead in Your Home*).

In general, most commenters were supportive of a requirement to distribute a new renovation-specific pamphlet for the purposes of TSCA section 406(b). One commenter stated a belief that the existing *Protect Your Family from Lead in Your Home* pamphlet had served its purpose well and the development of a new pamphlet should not be a priority. EPA agrees with the commenters who recognized the merit of providing renovation-specific information to owners and tenants before renovations commence. Therefore, this final rule requires renovation firms to distribute the *Renovate Right: Important Lead Hazard Information for Families, Child Care Providers and Schools* pamphlet before beginning renovations.

Information Distribution Requirements

Other than the use of the new renovation-specific pamphlet, EPA did not specifically propose any changes to the existing information distribution requirements for target housing that does not meet the proposed definition of "child-occupied facility." One commenter contended that the existing information distribution requirements for multifamily target housing were extremely burdensome and resulted in tenants being given multiple notifications and copies of the lead hazard information pamphlet over the course of a year's time. This commenter requested that EPA modify the regulations to allow an annual distribution of renovation-related lead hazard information to tenants. However, as noted in interpretive guidance previously issued on the Pre-Renovation Education Rule, EPA, in developing the final Pre-Renovation Education Rule, carefully weighed whether a one-time pamphlet distribution would be adequate to meet the objectives of section 406(b) of the lead statute, and concluded that many, if not most, tenants would benefit from receiving the information in the lead pamphlet closer to the time that a renovation is to begin.

Although some tenants may read lead information delivered on a "for-your-information" basis, many others are not likely to focus on potential lead hazards until a renovation affecting their unit is imminent, and would welcome receiving information on protecting their families from lead in a more timely fashion. Therefore, EPA has determined that an annual distribution of renovation-specific lead hazard information would not be an effective means of providing timely information to tenants.

However, with respect to renovations in common areas, EPA has determined that there are other effective ways of delivering lead hazard information to tenants in a timely manner. Specifically, the posting of informational signs during the renovation in places where the tenants of the affected units are likely to see them will provide these tenants with the information they need at the time that they need it. Depending on the circumstances, renovation firms may find the posting of such signs to be less burdensome than mailing or hand-delivering this information to affected tenants. Indeed sign posting may be more effective than mail since it provides an immediate reminder.

Therefore, EPA will allow renovation firms performing renovations in common areas of multi-unit target housing the option of mailing or hand-delivering general information about the renovation and making a copy of the pamphlet available to the tenants of affected units on request prior to the start of the renovation, or posting informational signs while the renovation is ongoing. These signs must be posted where they are likely to be seen by all of the tenants of the affected units and they must contain a description of the general nature and locations of the renovation and the anticipated completion date. The signs must be accompanied by a posted copy of the pamphlet or information on how interested tenants can review or obtain a copy of the pamphlet at no cost to the tenants.

One commenter expressed concern about tenants either not seeing the postings because they use different entrances or distinguishing the renovation-specific lead hazard information postings from other postings in the general area. To take advantage of this option, this final rule requires renovation firms to use actual signs, not notices on tenant bulletin boards. In addition, these signs must be posted where the tenants of all of the affected units can see them. If the tenants of the affected units use several different entrances, a sign posted by one of the entrances would not be sufficient.

With respect to renovations in individual housing units, whether single-family or multifamily, firms performing renovations for compensation in target housing must continue to distribute a lead hazard information pamphlet to the owners and tenants of the housing no more than 60 days before beginning renovations. This requirement, along with the associated requirements to obtain acknowledgments or document delivery, has not changed.

For renovations in the common areas of multi-unit target housing, firms must provide tenants with general information regarding the nature of the renovation and make the pamphlet available on request, by mailing, hand-delivery, or posting informational signs. Firms must also maintain documentation of compliance with these requirements. The 2007 Supplemental Proposal contained additional proposed information distribution requirements for COFs in target housing and in public and commercial buildings. This final rule incorporates those additional requirements.

Also, as proposed in the 2006 Proposal, this final rule deletes the existing 40 CFR 745.84 because it is duplicative. The section provided some details on submitting CBI and how EPA will handle that information. However, comprehensive regulations governing sensitive business information, including CBI under TSCA, are codified in 40 CFR, Part 2. The regulations in 40 CFR, Part 2, set forth the procedures for making a claim of confidentiality and describe the rules governing EPA's release of information. EPA received no comments on the proposed deletion of 40 CFR 745.84. Therefore, EPA is deleting this section and redesignating existing 40 CFR 45.85 as 40 CFR 745.84.

The Environmental Protection Agency is also taking this opportunity to reiterate who is responsible for complying with the information distribution responsibilities of 40 CFR 745.84. This provision of this final rule includes the existing Pre-Renovation Education Rule information distribution requirements as amended to include requirements applicable to child-occupied facilities. In interpretive guidance issued for the Pre-Renovation Education Rule, EPA shed additional light on the issue of who is responsible for complying with the information distribution requirements, particularly for renovation projects where multiple contractors are involved. EPA stated that if the renovation is overseen by a general contractor, the general contractor is considered to be the "renovator" under the rule and is therefore responsible for ensuring that the information distribution requirements are met.

EPA further stated that it would not consider a subcontractor to be a "renovator" for purposes of the Pre-Renovation Education Rule so long as the subcontractor has no direct contractual relationship with the property owner or manager relating to the particular renovation. EPA's reasoning is that the information distribution requirements should be fulfilled by the person or entity with which the customer enters into the contract and compensates for the work—even if that work is subsequently contracted out.

This final rule changes the existing definition of "renovator" to refer specifically to the individual trained in work practices as distinct from the renovation firm. The final rule also specifies in 40 CFR 745.84 that the renovation firm is responsible for carrying out the information distribution requirements. Renovation firms may find it more efficient to have someone other than the certified renovator distribute the pamphlet and obtain the acknowledgement forms. In changing the definition of "renovator," EPA is not changing its policies as to which entity, between a contractor and subcontractor, is responsible for carrying out the information distribution requirements. On the contrary, as to this issue, EPA intends to continue interpreting the regulatory responsibility for the information distribution requirements as it has in the past.

► COMMERCIAL BUILDINGS CONTAINING A CHILD-OCCUPIED FACILITY

Workers must provide a lead hazard information pamphlet to the owner of the building as well as to an adult representative of the child-occupied facility, if the owner of the building and the child-occupied facility are different entities. This requirement was modeled on the Pre-Renovation Education Rule's requirements for pamphlet distribution in rental target housing. As described in the 2007 Supplemental Proposal, EPA has determined, in accordance with TSCA, Section 407, that the distribution of lead hazard information, before renovation projects begin, to an adult representative of the COF as well as to the owners of public or commercial buildings that contain child-occupied facilities is necessary to ensure effective implementation of this regulation. EPA believes that information on lead-based paint hazards, and lead-safe work practices that minimize the creation of hazards, will stimulate interest on the part of child-occupied facilities and public or commercial building owners in these work practices and increase the demand for their use.

EPA received no comments on this aspect of the 2007 Supplemental Proposal. Therefore, the final rule includes this requirement as proposed. Renovation firms performing renovations for compensation in a COF in a public or commercial building must provide the lead hazard information pamphlet entitled *Renovate Right: Important Lead Hazard Information for Families, Child Care Providers and Schools* to the owner of the building. The renovation firm must either obtain written acknowledgment from the owner that the pamphlet was delivered or obtain a certificate of mailing for the pamphlet at least 7 days prior to the start of the renovation.

In addition, the renovation firm must provide the pamphlet to an adult representative of the COF if the facility and the building are owned by different entities. To document compliance with this requirement, the renovation firm must do one of the following:

- Obtain a written acknowledgment of pamphlet delivery from the adult representative of the COF
- Obtain a certificate of mailing for the pamphlet at least 7 days prior to the start of the renovation
- Certify in writing that the pamphlet has been delivered to the COF and the firm has been unsuccessful in attempting to obtain the signature of an adult representative of the child-occupied facility; this certification must contain the reason for the failure to obtain the signature

Parents and Guardians of Children under Age 6 Using a Child-Occupied Facility

The 2007 Supplemental Proposal would also have required that a renovation firm performing a renovation for compensation in a child-occupied facility provide information about the renovation to the parents and guardians of children under age 6 using the facility. This proposed requirement was designed to be comparable to the Pre-Renovation Education Rule provisions for informing adult occupants (who are not owners). EPA is finalizing this requirement as proposed. The renovation firm must either mail each parent or guardian the lead hazard information pamphlet and a general description of the renovation or post informational signs where parents and guardians would be likely to see them. The signs must be accompanied by a posted copy of the pamphlet or information on how to obtain the pamphlet at no charge to interested parents or guardians. This requirement applies to renovations in COFs in target housing as well as to renovations in child-occupied facilities in public or commercial buildings.

EPA received three comments on this aspect of the 2007 Supplemental Proposal. One commenter expressed support for this proposed requirement. The other two provided a number of reasons why the final rule should not include such a requirement. These commenters noted that renovation firms have no contractual connection with or contractual responsibility to the parents or guardians of children using a COF. They believe that the child-occupied facility owner bears primary responsibility for maintaining a safe environment for children.

They were also concerned that renovation firms might be called on to spend a significant amount of additional time at a COF to answer parents' questions about lead poisoning. EPA is not persuaded by these comments. Although the firms may have no contractual connection with the parents or guardians of the children, that is often the case with occupants who are not owners. Although owners of child-occupied facilities bear responsibility for maintaining a safe environment for children, renovation firms are responsible for providing the pamphlet to owners and occupants.

Once the renovation firm has distributed the pamphlet, it has no further obligation to educate the owners or occupants about lead poisoning. The pamphlet contains this information and refers to additional resources. EPA acknowledges that it may be difficult to provide copies of the pamphlet to each parent, which is why this final rule allows renovation firms to comply by posting informational signs where parents or guardians would be likely to see them.

In the next chapter we look at some of the needs for training and certification.

EPA Lead Training and Certification

7

Under the current Lead-Based Paint Activities Regulations at 40 CFR, Part 745, Subpart L, both individuals and firms that perform lead-based paint inspections, lead hazard screens, risk assessments, and abatements must be certified by the U.S. Environmental Protection Agency (EPA). EPA proposed a similar, but not identical, regulatory scheme for individuals and firms that perform renovations.

This final rule requires all renovations subject to this rule to be performed by a firm certified to perform renovations.

In addition, the rule requires that all persons performing renovation work either be certified renovators or receive on-the-job training (OJT) from and perform key tasks under the direction of a certified renovator. To become a certified renovator, a person must successfully complete an accredited renovator course. EPA renovator certification allows the certified individual to perform renovations in any state, territory, or Indian tribal area that does not have a renovation program authorized under 40 CFR, Part 745, Subpart Q. These requirements are discussed in greater detail in the following sections.

EPA is also creating, with this final rule, a dust sampling technician discipline. Although, as discussed in Unit III.E.7 of the preamble, this final rule does not allow dust clearance testing in lieu of post-renovation cleaning verification (except in limited circumstances), EPA still believes that there will be a market for the services of persons with dust sampling technician credentials. EPA recommends that any property owners who choose to have dust clearance testing performed after a renovation, use a certified inspector, risk assessor, or dust sampling technician.

Finally, in response to one commenter who suggested that EPA's use of the term "person" and the term "individual" was confusing, EPA has modified the regulatory text in the sections added or significantly revised by this final rule to use the term "person" when referring to both natural persons and judicial persons, such as renovation firms, property management companies, or units of government, and the term "individual" when referring only to natural persons.

Under this final rule, EPA is establishing new individual certification disciplines for renovators and dust sampling technicians. All renovation activities covered by this final rule must be performed by certified renovators, or by renovation workers who receive on-the-job training in the work practices from a certified renovator.

The certified renovator assigned to a renovation is responsible for ensuring that the renovation is performed in compliance with the work practice requirements set out in 40 CFR 745.85. These requirements pertain to warning signs and work area containment, the restriction or prohibition of certain practices (e.g., high heat gun, torch, power sanding), waste handling, cleaning, and post-renovation cleaning verification. The certified renovator can perform these work practices herself or himself. Alternatively, the certified renovator can direct other workers to perform most of these work practices. However, the post-renovation cleaning verification requirements must be performed by a certified renovator. These requirements cannot be delegated to a worker. If the certified renovator directs the other workers to perform the work practices, the certified renovator must be at the work site during the critical phases of the renovation activity. The critical phases are posting warning signs, containing the work area, and cleaning the work site.

Although the certified renovator is not required to be onsite at all times while the renovation project is ongoing, a certified renovator must nonetheless regularly direct the work being performed by other workers to ensure that the work practices are being followed. When a certified renovator is not physically present at the work site, the workers must be able to contact the renovator immediately by telephone or other mechanism. A certified renovator must:

- Perform the post-renovation cleaning verification described in 40 CFR 745.85(b)
- Perform or direct workers who perform all of the work practices described in 40 CFR 745.85(a)
- Provide training to workers on the work practices they will be using in performing their assigned tasks

- Be physically present at the work site when the signs required by 40 CFR 745.85(a)(1) are posted, while the work area containment required by 40 CFR 745.85(a)(2) is being established, and while the work area cleaning required by 40 CFR 745.85(a)(5) is performed
- Regularly direct the work being performed by other workers to ensure that the work practices are being followed, including maintaining the integrity of the containment barriers and ensuring that dust or debris does not spread beyond the work area
- Be available, either onsite or by telephone, at all times that renovations are being conducted
- When requested by the party contracting for renovation services, use an acceptable test kit to determine whether components to be affected by the renovation contain lead-based paint
- Have with them at the work site copies of their initial course completion certificate and their most recent refresher course completion certificate
- Prepare the records required to demonstrate that renovations have been performed in accordance with the requirements of this rule

There are some slight revisions between the 2006 Proposal and this final rule, although none of these changes add to or detract from the renovator's responsibilities. First, the Proposal used both the term "lead-safe work practices" and "work practices" in the preamble and in the proposed rule text. Although the work practices required in this final rule are lead-safe, for purposes of clarity, the final rule text has been changed to "work practices." The reason for this change was to make text of the rule relating the renovator's responsibilities consistent with other provisions in the rule, particularly 40 CFR 745.85 (Work Practice Standards).

Today's work practices are lead-safe work practices. The work practice standards listed in §745.85(a) are the same tasks that the other workers will be directed in and trained to do by the certified renovator (except for cleaning verification). In addition, the term *lead-safe work practices* has different meanings in different contexts, and this change is to make clear that the work practices required by this final rule are the work practices required in §745.85(a).

Second, one of the renovator's responsibilities listed in the preamble of the 2006 Proposal was to "regularly direct the work being performed by uncertified persons to ensure that lead-safe work practices are being followed, the integrity of the containment barriers is maintained, and dust or debris is not spread beyond the work area."

The word "regularly" was inadvertently omitted from the proposed regulatory text. To make the regulatory text consistent with the preamble, the word "regularly" has been added to the final regulatory text. In addition, EPA has slightly modified the regulatory text, consistent with the preceding paragraph, to clarify that maintaining the integrity of the containment barriers and ensuring that dust or debris does not spread beyond the work area are among the work practices required by the rule.

Some commenters agreed that it was unnecessary for a certified renovator to be onsite at all times and believed that oversight by a certified renovator on a regular basis was sufficient. One commenter believed that the certified renovator should be onsite at critical points including site preparations and isolation, end of day and end of project cleaning, and cleaning verification. Many other commenters thought a certified renovator should be onsite at all times. Another stated that a certified renovator would not have to be onsite at all times if workers received lead-safe work practices training. After carefully considering the issue, EPA has concluded that requiring a certified renovator to be onsite during critical phases of the work is sufficient to ensure that the work practices required by this final rule are followed.

These work practices provide a mechanism to contain dust and debris generated by a job and a cleanup regimen following work that is designed to minimize exposure to lead-based paint hazards created during the renovation activity. Once the containment has been established and until cleanup begins, this final rule requires few, and simple, changes from the way renovation work is currently carried out. Specifically, renovation workers need to avoid using the specific practices prohibited by this final rule; they need to maintain the containment (e.g., avoid ripping or displacing the plastic); and they need to make sure that any waste generated is contained at the end of the day. These are important but relatively simple measures that EPA does not believe require formal classroom training, or the constant supervision of a certified renovator who has had formal training.

Once the cleanup begins, the certified renovator will again be required to be present, either performing the cleanup or directing others. In addition, the certified renovator must perform the cleaning verification. Thus, EPA has concluded that having a renovator onsite at all times is unwarranted.

▶ RENOVATOR TRAINING

To become a certified renovator, a person must successfully complete a renovator course accredited by EPA or by a state, territorial, or tribal program authorized by EPA.

Some commenters questioned the need to create a separate discipline for renovators. In their opinion, the existing abatement course is sufficient (with some basic changes) and to create a new program will take resources away from existing efforts in lead hazard control. EPA believes that there are sufficient differences between abatement and renovation activities to warrant different training and work practice requirements. Specific activities of an abatement contractor may be similar to those of a renovator (e.g., sanding, caulking, painting, sawing), but because the project goal is the permanent elimination of hazards, the application and methodology differ.

Therefore, a significant portion of an abatement contractor's training is focused on abatement techniques and selection of the appropriate course of action for a variety of hazards. Renovators, on the other hand, do not seek to permanently eliminate lead hazards. Renovators perform maintenance and improvement tasks as directed by the consumer. The goal of EPA's renovator training and certification program is not to update the methodology a renovator uses to accomplish these tasks, with the exception of the practices prohibited or restricted by this final rule, but rather to introduce containment and cleaning methods to minimize exposure to lead-based paint hazards created by the renovation activity.

Several commenters saw the need for universal, standard renovator training. A commenter suggested that training for certified renovators be similar to the current EPA/HUD renovator and remodeler course. One commenter thought that standard training would make it easier when hiring someone to verify that they had completed the appropriate training. Another mentioned that it would encourage state-to-state reciprocity for training programs so that renovators would not need to take multiple courses with the same content.

EPA plans to work with HUD to update the model EPA/ HUD renovator training course to cover the requirements of this final rule. EPA agrees that reciprocity among authorized state, territorial, and tribal programs, and with the federal program, is preferable. However, as with the abatement program, authorized programs will have the ability to customize requirements and course content based on their particular needs. The agency encourages jurisdictions seeking authorization to consider reciprocity of training as they develop their individual programs. Appropriate notice must be provided for training courses. This includes both training notification and post-training notification.

Commenters were also concerned about the cost of formal training, suggesting that EPA could provide free training to encourage renovator compliance, or that EPA funds for enforcement of the final rule would be better spent on training. EPA agrees that renovator training should be as inexpensive as possible. However, the training course costs will be established by independent training programs based on market forces. The total cost of conducting a training course depends on the labor cost for the instructor(s), the cost of providing a classroom and other facilities, and other fixed costs. But the cost per trainee also depends on the number of trainees per class. Due to the large number of individuals who will need training, the agency anticipates that demand will be high, keeping the cost per trainee lower than might otherwise be the case. But also due to that large volume, the agency does not anticipate that it will be able to provide any significant source of funding to support training.

Training for Other Renovation Workers

This final rule does not require everyone involved in performing a regulated renovation project to receive training from an accredited training provider. To allow flexibility for firms undertaking these projects, the rule allows firms to use other workers to perform renovation activities as long as they receive OJT in work practices from a certified renovator. This training must include instruction in the specific work practices that these workers will be responsible for performing.

OJT training occurs while the worker is engaged in productive work and provides knowledge and skills essential to the full and adequate performance of the job. OJT may also be structured through a planned process of developing competence on units of work by having the certified renovator train the worker at the work setting or a location that closely resembles the work setting. Although there is no specific requirement for "refresher training," OJT must be provided for each worker for each job to the extent necessary to ensure that that worker is adequately trained for the tasks he or she will be performing.

If, under the direction of the certified renovator, the workers will be posting warning signs, establishing containment, or cleaning the work area after the renovation, the certified renovator must provide instruction, either verbally or through demonstration, to the workers in how to perform these tasks. With respect to other activities, including work performed while the certified renovator is not present, the certified renovator must provide instruction, either verbally or through demonstration, in how to perform the work without using work practices prohibited by this rule, how to maintain the integrity of the containment barriers (e.g., taking care not to tear the plastic), and how to avoid spreading dust or debris beyond the work area (e.g., vacuuming clothing and tools with a HEPA vacuum before leaving the work area). In any event, the certified renovator remains responsible for ensuring that this work is done in compliance with the rule's requirements, for example, that containment sufficient to prevent release of dust or debris from the work site has been established and that clothing and tools were adequately cleaned before leaving the work area.

Workers need not be trained in work practices that do not pertain to the renovations they will be performing. If the certified renovator will be the one posting warning signs, establishing containment, and cleaning the work area after the renovation, it is not necessary for the certified renovator to provide instruction on these tasks to any workers who will be used elsewhere on the project. Similarly, workers hired to perform only exterior projects need not receive training in how to clean an interior work area after a renovation.

The Environmental Protection Agency chose to allow OJT to alleviate industry concerns raised during the Small Business Regulatory Enforcement Act (SBREFA) panel process regarding high employee turnover rates within the industry and the potential for high training costs if all workers were required to be certified. The agency concluded that OJT could be done effectively and would provide flexibility for firms undertaking renovation projects. EPA determined that OJT can be effectively delivered by a certified renovator because the requirements themselves are simple and easy to understand. This final rule also requires a certified renovator be assigned and responsible for each project to ensure compliance with required standards.

Some commenters agreed that OJT by a certified renovator is sufficient for training workers. One commenter stated that as long as a specific person is designated to oversee the job, there is no need for all workers onsite to have formal training. The commenter noted the similarity between this approach and OSHA's "competent person" standard. EPA agrees that there are some similarities between the approach in this final rule and OSHA's "competent person" standard.

However, the majority of commenters had concerns about the use of OJT to train workers. Many argued that OJT is insufficient for providing workers with the necessary skills and thought renovation workers should receive formal Lead-Safe Work Practices (LSWP) training such as a 1-day course equivalent to that required for certified renovators. Some of these commenters also thought that workers should be certified or licensed.

Some commenters were concerned that the content of OJT is not clearly defined in the rule. One believed EPA should impose a structured OJT program to produce consistent, accurate, and comprehensive training outcomes. Others thought more time was needed for OJT, with suggestions ranging from 5 to 6 hours of training to 3 to 4 days. EPA has neither established a structured OJT program nor required a specific length of time for OJT because the OJT required will vary widely from project to project, depending upon how the other workers are used. As discussed above, if the worker will not be establishing containment, there is no need to train the worker in how to establish containment. If the worker in question is an electrician, and he or she will merely be installing an electrical outlet as part of a larger job, then there may be no need to provide any training to this worker other than instructing him or her not to disturb the plastic on the floor and making sure that the tools are free of dust and debris before leaving the work area.

EPA will "grandfather" persons with previous EPA/HUD lead-safe work practices training or accredited abatement supervisor or worker training. To become certified renovators, these persons must take a renovator refresher course to ensure they are acquainted with how to use test kits to determine whether lead-based paint is present on a component and how to perform cleaning verification. However, even if they do not take the refresher course and become certified renovators, these individuals have still received significant training in the required work practices such as establishing containment and cleaning the area after the job is finished. They are not likely to need much, if any OJT, depending on how recent their training was. Similarly, although not recognized for the purpose of "grandfathering" by EPA, HUD's Lead Maintenance course would also provide a great deal of information on lead-safe work practices. Someone who had taken the Lead Maintenance course recently would also not be likely to need much, if any, OJT.

Several commenters thought that workers would not receive adequate OJT because the certified renovator was not qualified to train others. They noted that the certified renovators are renovators, not professional trainers, and do not necessarily have the skills necessary for teaching others.

After consideration of these commenters' concerns, EPA has concluded that OJT is sufficient for training some renovation employees. The work practice standards of this final rule are not complex or difficult to institute, and those activities critical to ensuring the lead-safe outcome of the project are either conducted by certified renovators or directed by certified renovators. The remainder of the project is often just the renovation itself, and EPA was careful when developing these final work practices to minimize the effect on the way typical renovations are conducted. With the exception of the prohibition of certain unsafe practices, renovation methods are unaffected by this rule. For example, the work practices of this final rule do not affect the method a firm would employ to replace a window.

A certified renovator should be able to demonstrate to other firm employees work practices, such as how to work within containment and how to move into and out of containment without spreading lead dust and debris. EPA does not believe a professional trainer is needed to train renovation workers, who will be directed by a certified renovator if they will be performing any of the key tasks associated with the work practices. Most of the people performing renovations today are not trained by professional trainers. They are trained on the job by experienced firm employees. For example, persons learn the various techniques for removing and replacing windows from others in the firm who are experienced in these techniques. Renovation workers can learn work practices in the same way from a certified renovator.

Although the work practices in the final regulation are sufficiently straightforward and can be easily demonstrated by the certified renovator, EPA agrees that renovators do not necessarily consider themselves to be trainers. Therefore, accredited renovator training will include a train-the-trainer component to provide instruction on providing OJT. In addition, instructors will be expected to provide training tips to renovators during hands-on instruction. As the instructor is showing the renovator how to do these work practices, he or she can also provide instruction on how to show others how to do these work practices. Accordingly, EPA has concluded that certified renovators will be adequately prepared to provide OJT that is sufficient and appropriate for the purposes of this rule.

Commenters expressed concerns that the rule would not provide appropriate training for the large number of non-English speaking workers in the renovation field. One of these commenters suggested that EPA consider such means as graphic manuals, video presentations, and translators to aid in training non-English speaking workers. Another thought that a hands-on-only training process overlooked possible language barriers between the certified renovator and trainee. EPA agrees that OJT can be conducted effectively by demonstration by the certified renovator or through the use of graphic training materials.

The agency plans to develop materials to assist certified renovators in conducting on-the-job training. To the extent possible, these materials will use a graphic format that does not require the use of any particular language. Moreover, renovation firms currently communicate job needs to their employees, and EPA doubts that firms routinely hire people with whom they are unable to communicate. Finally, EPA emphasizes again that the certified renovator and the renovation firm are responsible for ensuring compliance with this final rule. If the certified renovator has doubts about an employee's understanding of or ability to comply with the requirements that are relevant to the work he or she is to undertake, the certified renovator may need to be onsite and direct the work more regularly than he or she otherwise would, or may need to perform certain tasks him or herself. However, given the relative simplicity of the work practices that are required between establishment of containment and cleanup, EPA does not expect that this will often be necessary.

Some commenters were concerned that OJT does not include a means to assess worker competence such as an examination. Commenters were also concerned about ongoing training needs and suggested requiring worker refresher training on a periodic or annual basis. This final rule requires a certified renovator to direct workers with OJT as necessary to ensure that work practices are being followed. This will necessarily involve a period of observation after OJT is provided to ensure that the worker has understood and is following the work practices pertinent to his assigned duties.

In addition, to some extent, OJT is continuous and certified renovators will likely need to continue to provide training to workers based on the activities that they will be expected to perform on a particular job. A certified renovator would not need to provide OJT to the same worker on consecutive jobs if the worker is performing the same work, but if the nature of the work varies, or if the firm hires a new employee, relevant OJT would have to be provided for the work to be performed. EPA believes that the continuous nature of OJT obviates the need for a refresher training requirement in the rule and will serve as an incentive for firms to have their permanent employees trained as certified renovators. EPA also believes that refresher training *per se* is not practical, given that OJT will be specific to the job in question.

Some commenters wanted some form of verification that a worker had received training, such as a certificate of training or a sticker that could be placed on an ID card. Because each worker is not likely to receive training in all aspects of lead-safe work practices, a certificate or other form of training completion that would indicate that an employee's OJT is complete is not appropriate for this program.

It is important to note that OJT is not as portable as certified renovator training nor is it intended to be. Certified renovators carry a training certificate that they can present to each new employer to prove that they have received training in the required work practices. There is no corresponding document that can be used to verify OJT by a previous employer. Renovation firms will generally need to provide OJT each time a new worker is used. It is also the renovation firm's responsibility to adequately document the elements of OJT provided to each worker on each project.

Because a certified renovator must be assigned to each and every renovation covered by this regulation, EPA anticipates that some renovation contractors and property management companies will find that they achieve maximum efficiency and flexibility by qualifying all of their permanent employees who perform renovations as certified renovators. However, as a result of the industry's high employee turnover rates and short-term labor needs, the agency believes that training flexibility in the form of OJT is needed. EPA believes that such flexibility will provide firms the ability to respond to variable labor demands and will not compromise the safety of this final rule. EPA is concerned that a regulation requiring formal, classroom training for every worker performing any renovation activity would be unrealistic for this industry and therefore less effective at ensuring that the renovation workforce is trained in work practices than the more balanced training requirements in this final rule.

Dust Sampling Technicians

Except as provided in 40 CFR 745.85(c), this final rule does not allow dust clearance sampling to be performed in lieu of post-renovation cleaning verification. However, some property owners may still choose to have dust clearance sampling performed after the renovation. Dust sampling technicians certified in accordance with this final rule will be available to perform dust clearance sampling after renovations and for the purposes of HUD's Lead-Safe Housing Rule.

Some commenters questioned the need for dust sampling technicians. One stated that there is no benefit to creating a third inspection-type discipline that has such limited training requirements. Two commenters thought that only EPA- or state-certified risk assessors should be allowed to collect dust wipe clearance samples and two commenters thought that dust sampling technicians should be required to work under a certified risk assessor or inspector.

In 1999, to make accurate dust testing for lead more available and affordable, Congress provided EPA with funding for the development of a 1-day dust sampling technician course. Congress also encouraged the agency to promote the recognition of this discipline. EPA completed the development of this course, entitled "Lead Sampling Technician Training Course," in July 2000. This course provides instruction on how to conduct a visual assessment for deteriorated paint, collect samples for lead dust, and interpret sample results. The training curriculum provides clearance sampling instruction equivalent to that presented in inspector and risk assessor courses, in terms of time and quality with respect to dust sampling. Therefore, EPA can recommend that property owners and others who wish to have optional dust sampling performed use the services of a certified inspector, risk assessor, or dust sampling technician.

▶ INITIAL CERTIFICATION OF INDIVIDUALS

Section 745.90 of this final rule addresses renovator and dust sampling technician certification. To become a certified renovator, a person must successfully complete a renovator course accredited by EPA or by a state, territorial, or tribal program authorized by EPA under 40 CFR, Part 745, Subpart Q. The renovator course accreditation requirements are based on the joint EPA–HUD model curriculum entitled "Lead Safety for Remodeling, Repair, & Painting." EPA is not requiring additional education or work experience of persons wishing to become certified renovators.

EPA renovator certification will allow the certified individual to perform renovations covered by this section in any state or Indian tribal area that does not have a renovation program authorized under 40 CFR, Part 745, Subpart Q. To become a certified dust sampling technician, a person must successfully complete a dust sampling technician training course that has been accredited either by EPA or by a state, territorial, or tribal program authorized by EPA under 40 CFR, Part 745, Subpart Q. EPA is not requiring additional education or work experience of persons wishing to become certified dust sampling technicians.

The final rule also establishes, in 40 CFR 745.91, procedures for suspending, revoking, or modifying an individual's or firm's certification. These procedures are very similar to the current procedures in place at 40 CFR 745.226(i) for suspending, revoking, or modifying the certification of an individual who is certified to perform lead-based paint activities. In addition, under the final rule, renovator certification can be suspended, revoked, or modified if the certified renovator does not conduct projects to which he or she is assigned in accordance with the work practice requirements of this final rule.

Finally, to ensure that the effect of a suspension, revocation, or modification determination is clear to the certified individual or firm, EPA has added language to this section ensuring that the commencement date and duration of a suspension, revocation, or modification is identified in the presiding officer's decision and order. EPA has also added language to this section to clarify what steps an individual or firm must take after such an action to exercise the privileges of certification again.

An individual whose certification has been suspended must take a refresher training course in the appropriate discipline to make his or her certification current, while an individual whose certification has been revoked must take another initial training course to be recertified. A firm whose certification has been suspended need not do anything after the suspension ends to become current again, as long as the suspension ends before the firm's certification expires. If the firm's certificate expires during the suspension, the firm must apply for recertification after the suspension ends. If a firm's certification is revoked, the firm must apply for certification after the revocation period ends to be certified.

Some commenters questioned the need for a certification requirement, emphasizing that it is the training that is important rather than the certification. One commenter thought that, since firms will have to be certified, there was no added value in certifying renovators. Others supported certification and some thought renovators should have to apply to EPA to receive their certification in the same way that abatement workers do, stating that no regulatory program can work unless the regulating agency can reliably identify and contact the regulated individuals. One commenter thought that there should also be a work experience requirement for certified renovators.

EPA believes that renovators must be certified so that the agency has a mechanism to verify an individual has received the appropriate training. In addition, if a contractor does not comply with the regulatory standards then withdrawal of the renovator's certification is a regulatory remedy available to the agency. The final rule includes a certification process that is more streamlined than the individual certification process of the agency's abatement regulations. In the abatement program, an individual must complete training, then submit an application and fee to the agency and, depending on the discipline, take a third-party exam to be certified. In contrast, an individual will be considered a certified renovator on successful completion of an accredited training program, and the accredited training program is required to submit identifying and contact information to EPA regarding the individuals that they have trained.

EPA does not believe that work experience requirements are necessary because previous experience in the construction or renovation industry would do little to help an individual understand or perform the work practices, which are not standard practice in the industry. Consequently, there is no relevant work experience for EPA to require. In addition, the work practices required by this final rule are sufficiently straightforward that EPA does not believe it is necessary to require work experience in addition to certified renovator training.

Because EPA is not requiring any additional education or work experience requirements, or a third-party examination similar to that taken by inspector, risk assessor, or supervisor candidates, EPA believes that there is little value in requiring candidates to apply to EPA to receive their renovator or dust sampling technician certification. Currently, the only certified discipline without prerequisites in education or experience, or a third-party examination, is the abatement worker. When candidates for worker certification apply to EPA, EPA verifies that the copy of the training course certificate submitted with the application is from an accredited training provider. Without requiring renovators or dust sampling technicians to apply to EPA for certification, EPA will still receive course completion information from course providers. With this information, EPA will have a complete list of certified renovators and will be able to check to see if a particular course completion certificate holder appeared on a course completion list submitted by the training course provider identified on the certificate.

When EPA inspects a renovation job for compliance with these regulations, EPA will have the ability to verify, to the same extent, the validity of a course completion certificate held by a renovator at that job. Therefore, under this final rule, EPA is requiring that a course completion certificate from an accredited training provider serve as a renovator's or dust sampling technician's certification. To facilitate compliance monitoring, the rule requires a certified renovator or dust sampling technician to have a copy of the course completion certificate at the job site.

Several commenters saw the need for a way to determine that a certified renovator was current with applicable training requirements. Suggestions for proof of training included issuing photo IDs, issuing a hard card or certificate, and establishing a national database of workers with current training. One commenter thought that it should be the responsibility of the training provider to certify that renovators have successfully completed the training requirements and to then supply EPA with all of the information. EPA agrees that there must be a way to determine if a renovator is certified and is current with training requirements.

The agency agrees that a database of renovator information would be important, and will include identifying and training information in the agency's Federal Lead Paint Program (FLPP) database. However, this database will only contain information about certified renovators working in federally administered jurisdictions. In addition, the agency will require training programs to include a photograph of the individual who completes renovator or dust sampling technician training on the training certificate and to submit that photo to the agency to be included in the database record. This will enable inspectors to determine whether a particular individual has received training from an accredited training provider.

Some of the commenters had concerns specific to small businesses. Two commenters stressed the need for outreach programs to inform small businesses of new compliance requirements. One commenter stated that smaller firms should not be exempt from training and certification requirements; another thought that small businesses would continue to operate without appropriate training and certification unless there was some type of enforcement. EPA understands that the task of communicating this final rule requirement to the renovation community will be challenging. Therefore, EPA is developing a comprehensive out-reach and communications program to support this final rule. This will include outreach to contractors as well as consumers. In addition the agency plans to roll out a compliance assistance effort to complement this undertaking.

One commenter suggested that authorized state, territorial, or tribal programs include the requirement for training as part of a contractor licensing function, thereby eliminating the need to create a special (new) lead renovator's certification or license. EPA agrees that where a state, territory, or tribe has a preexisting relationship with renovation contactors, such as a renovators' licensing program, the simplest and most cost-effective approach may be to incorporate a requirement for lead-safe work practice training into that preexisting program.

Recertification

Under this final rule EPA is requiring that renovators and dust sampling technicians who wish to remain certified take refresher training every 5 years. In addition, EPA is requiring that the refresher training course be half the length of the initial course. This is consistent with current practice for certified individuals performing lead-based paint activities. If an individual does not take a refresher course within 5 years of the date he or she completed the initial course or the previous refresher course, that individual's certification will expire on that date and that individual may no longer serve as a certified renovator or dust sampling technician. There is no grace period. To become cer-tified again, the individual must take another initial training course. In addition, under this final rule a certified renovator may choose to take the initial renovator course instead of a refresher course to allow maximum flexibility, particularly if, for some reason, the person was unable to attend a refresher course.

Some commenters asserted that the refresher requirement was of no benefit or imposed an unnecessary cost. These commenters reasoned that lead-safe work practices were not likely to change significantly over time. One noted that HUD's experience with lead-safe work practices training since 1999 has not revealed a need for refresher training in their program. Commenters who supported refresher training differed on the frequency of the training and the length of the refresher course. Some agreed that refresher training should be required every 3 years, others thought it should be required biennially, annually, or every 3 to 6 months.

One commenter agreed with the proposed 4-hour course, two commenters thought a 4-hour course was too short, and one thought that instead of completing a refresher, certified renovators should be required to retake the initial training course every 2 to 3 years. One commenter stated that a certified renovator should have the opportunity to take a third-party test and allow the renovator to "test out" of having to complete the refresher course.

After considering the range of concerns raised by the commenters, EPA has concluded that refresher training is important for renovators and dust sampling technicians and for the agency. During the refresher course, renovators and dust sampling technicians are given the opportunity to discuss any point of emphasis and to be updated on changes in the regulations or technical issues. For example, refresher training could be used to update renovators on availability of new techniques and products, such as test kits. Refresher training provides the agency with a mechanism to pass along critical information to certified individuals and to keep track of the workforce. However, EPA has determined that these purposes can be adequately served by 4-hour refresher training every 5 years, instead of every 3 years. This provides a reasonable period between trainings that limits training costs while providing an opportunity to update renovators and dust sampling technicians regarding regulations and technical issues. EPA believes that most renovators will not also be certified abatement professionals, so the difference in the length of time between required refresher courses should not confuse individuals about their responsibilities under the two programs.

Grandfathering

Under this final rule, individuals who successfully completed an accredited abatement worker or supervisor course, and individuals who successfully completed either HUD, EPA, or the joint EPA/HUD model renovation training courses may take an accredited refresher renovation training course in lieu of the initial renovation training to become a certified renovator. In addition, individuals who have successfully completed an accredited lead-based paint inspector or risk assessor course, but are not currently certified in the discipline, may take an accredited refresher dust sampling technician course in lieu of the initial training to become a certified dust sampling technician. Inspectors and risk assessors who are certified by EPA or an authorized program are qualified to perform dust sampling as part of lead hazard screens, risk assessments, or abatements. Therefore, it would be unnecessary for a certified inspector or risk assessor to seek certification as a dust sampling technician.

A number of commenters thought that certification should be given to those who have already attended appropriate training. Some of these commenters thought that individuals who had received EPA, HUD, or state-approved Lead-Safe Work Practices (LSWP) training should be grandfathered. One commenter thought individuals who had completed the OSHA 40-hour Hazardous Waste Operations and Emergency Response course should also be grandfathered and another wanted individuals who had taken the National Apartment Association's lead worker training course to be grandfathered. Four commenters were in favor of grandfathering dust sampling technicians who have previously completed a dust sampling course.

Most of the commenters who expressed an opinion agreed with grandfathering previously trained individuals but suggested that there be restrictions. Some of these commenters thought that to receive credit the training needed to have been completed in the last 2 to 3 years while others thought that certification should be given only if a refresher or "gap" course were completed. One commenter thought that the quality of the previous course should be taken into account and another commenter thought that a one-size fits all rule would not be appropriate and that factors including previous course requirements, the facility that had provided the training, and time elapsed since initial training should all be considered in establishing requirements for streamlined certification. One commenter opposed grandfathering, noting that existing courses do not cover lead test kits, cleaning verification, or recordkeeping in accordance with the proposed rule.

The final rule allows individuals who have successfully completed model renovation courses developed by HUD or EPA and individuals who have taken an abatement worker or supervisor course accredited by EPA or an authorized state or tribal program to become certified renovators by taking EPA-accredited renovator refresher training. Individuals who have successfully completed a risk assessor or inspector course accredited by EPA or an authorized state or tribal program can become certified dust sampling technicians by taking EPA-accredited dust sampling technician refresher training. EPA is recognizing only EPA and HUD model renovation training and lead-based paint activities training courses accredited by EPA or an authorized state, territorial, or tribal program because EPA has not sufficiently evaluated the content of other courses. In addition, it would be unwieldy to develop the content of multiple refresher courses based on the content of different initial training courses.

While the recognized training provides meaningful information relevant to these disciplines, it does not include some specific requirements of this final regulation. Therefore, EPA is requiring these individuals to receive refresher training to ensure they are familiar with the requirements of this final rule. Training providers are required to notify EPA of the individuals who become certified by successfully completing the refresher training. This information will support EPA's compliance assistance programs.

▶ RESPONSIBILITIES OF RENOVATION FIRMS

Under this final rule, firms must ensure that all persons performing renovation activities on behalf of the firm are either certified renovators or have been trained and are directed by a certified renovator in accordance with 40 CFR 745.90. The firm is responsible for assigning a certified renovator to each renovation performed by the firm and ensuring that the certified renovator discharges all of the responsibilities identified in this final rule. The firm must ensure that the information distribution requirements in 40 CFR 745.84 are met. As mentioned above, the certified renovator is responsible for ensuring compliance with 40 CFR 745.85 at all renovations to which he or she is assigned. The firm is also responsible for ensuring that all renovations performed by the firm are performed using certified renovators and in accordance with the work practice standards in proposed 40 CFR 745.85.

Where multiple contractors are involved in a renovation, any contractor who disturbs, or whose employees disturb, paint in excess of the minor maintenance exception is responsible for compliance with all of the requirements of this final rule. In this situation, renovation firms may find it advantageous to decide among themselves which firm will provide pre-renovation education to the owners and occupants, which firm will establish containment, and which firm will perform the post-renovation cleaning and cleaning verification. For example, a general contractor may be hired to conduct a multifaceted project involving the large-scale disturbance of paint, which the general contractor then divides up among several subcontractors. In this situation, having the general contractor discharge the obligations of the Pre-Renovation Education Rule is likely to be the most efficient approach, since this only needs to be done once.

With regard to containment, the general contractor may decide that it is most cost-effective to establish one large work area for the entire project. In this case, from the time that containment is established until post-renovation cleaning verification occurs, all general contractor and subcontractor personnel performing renovation tasks within the work area must be certified renovators or trained and directed by certified renovators in accordance with this rule. In addition, these personnel are responsible for ensuring the integrity of the containment barriers.

The cleaning and post-renovation cleaning verification could be performed by any properly qualified individuals, without regard to whether they are employees of the general contractor or a subcontractor. However, all contractors involved in the disturbance of lead-based paint, or who perform work within the work area established for the containment of lead dust and debris, are responsible for compliance with this final rule, regardless of any agreements the contractors may have made among themselves.

Initial Certification of Firms

This final rule requires firms that perform renovations, as defined by this rule, to be certified by EPA. EPA is adding a definition of *firm* to §745.83 to make it clear that this term includes persons in business for themselves (i.e., sole proprietorships), as well as federal, state, tribal, and local governmental agencies, and nonprofit organizations. Firms covered by this final rule include firms that typically perform renovations, such as building contractors or home improvement contractors, as well as property management companies or owners of multifamily housing performing property maintenance activities that include renovations within the scope of this final rule.

This final rule provides information about the certification and recertification process, establishes procedures for amending and transferring certifications, and identifies clear deadlines. A firm wishing to become certified to perform renovations must submit a complete "Application for Firms," signed by an authorized agent of the firm, along with the correct certification fee. EPA intends to establish firm certification fees in a separate rulemaking. EPA will approve a firm's initial application within 90 days of receipt if it is complete, including the proper amount of fees, and if EPA determines that the environmental compliance history of the firm, its principals, or its key employees does not show an unwillingness or inability to comply with applicable environmental statutes or regulations.

EPA will generally consider the following to be an indication that the applicant is unwilling or unable to comply with environmental statutes or regulations if, during the past 3 years, the applicant has:

- A criminal conviction under a federal environmental statute
- An administrative or civil judgment against the applicant for a willful violation of a federal environmental statutory or regulatory requirement
- More than one administrative or civil judgment for a violation of a federal environmental statute; violations that involve only recordkeeping requirements will not be considered

If the application is approved, EPA will establish the firm's certification expiration date at 5 years from the date of EPA's approval.

EPA certification will allow the firm to perform renovations covered by this section in any state or Indian tribal area that does not have a renovation program authorized under 40 CFR, Part 745, Subpart Q. If the application is incomplete, EPA will notify the firm within 90 days of receipt that its application was incomplete, and ask the firm to supplement its application within 30 days. If the firm does not supplement its application within that period of time, or if EPA's check into the compliance history of the firm revealed an unwillingness or inability to comply with environmental statutes or regulations, EPA will not approve the application and will provide the applicant with the reasons for not approving the application. EPA will not refund the application fees. A firm could reapply for certification at any time by filing a new, complete application that included the correct amount of fees.

This final rule provides firms with more time to amend their certification whenever a change occurs. A firm must amend its certification within 90 days whenever a change occurs to information included in the firm's most recent application. If the firm failed to amend its certification within 90 days of the date the change occurred, the firm would not be authorized to perform renovations until its certification was amended. Examples of amendments include a change in the firm's name without transfer of ownership, or a change of address or other contact information. To amend its certification, a firm must submit an application, noting on the form that it was submitted as an amendment. The firm must complete the sections of the application pertaining to the new information, and sign and date the form. The amendment must include the correct amount of fees.

Amending a certification will not affect the validity of the existing certification or extend the certification expiration date. EPA will issue the firm a new certificate if necessary to reflect information included in the amendment. Firm certifications are not transferable; if the firm is sold, the new owner must submit a new initial application for certification in accordance with 40 CFR 745.89(a). The final rule also includes procedures for suspending, revoking, or modifying a firm's certification. These procedures are very similar to the current procedures in place for suspending, revoking, or modifying the certification of a firm that is certified to perform lead-based paint activities.

Some commenters questioned the need for firm certification, while others, including industry representatives, supported it. The agency believes that firm certification is necessary for several reasons. First, certification is an important tool for the agency's enforcement program. To become certified, a firm acknowledges its responsibility to use appropriately trained and certified employees and follow the work practice standards set forth in the final rule. This is especially important under this final rule, since the certified renovator is not required to perform or be present during all of the renovation activities. Under these circumstances, it is important for the firm to acknowledge its legal responsibility for compliance with all of the final rule requirements, since the firm both hires and exercises supervisory control over all of its employees. Should the firm be found to violate any requirements, its certification can be revoked, giving the firm a strong incentive to ensure compliance by all employees.

Recertification

Under 40 CFR 745.89(b), a certified firm maintains its certification by submitting a complete and timely "Application for Firms," noting that it is an application for recertification, and paying the required recertification fee. With regard to the timeliness of the application for recertification, if a complete application, including the proper fee, is postmarked 90 days or more before the date the firm's current certification expires, the application will be considered timely and sufficient, and the firm's existing certification will remain in effect until its expiration date or until EPA has made a final decision to approve the recertification application, or not, whichever occurs later. If the firm submits a complete recertification application fewer than 90 days before the date the firm's current certification expired, EPA might be able to process the application and recertify the applicant before the expiration date, but this would not be guaranteed. If EPA does not approve the recertification application before the existing application expires, the firm's certification expires and the firm is not able to conduct renovations until EPA approves its recertification application. In any case, the firm's new certification expiration date will be 5 years from the date the existing certification expired.

If the firm submits an incomplete application for recertification and EPA does not receive all of the required information and fees before the date the firm's current certification expires, or if the firm does not submit its application until after its certification expired, EPA will not approve the firm's recertification application. The firm cannot cure any deficiencies in its application package by postmarking missing information or fees by its certification expiration date. All required information and fees must be in EPA's possession as of the expiration date for EPA to approve the application. If EPA does not approve the application, the agency will provide the applicant with the reasons for not approving the recertification application. Any fees submitted by the applicant will not be refunded, but the firm can submit a new application for certification, along with the correct amount of fees, at any time.

As with initial applications, this final rule includes a description of the actions EPA may take in response to an application for recertification and the reasons why EPA will take a particular action. This section is identical to the process for initial applications, except that EPA will not require an incomplete application to be supplemented within 30 days of the date EPA requests additional information or fees. In the recertification context, the firm must make its application complete by the date that its current certification expires.

Several commenters thought that firms should not be required to be recertified because the firm's certification is not based on knowledge or technology, but rather on a promise to abide by the rules. The agency believes that firm recertification is an important element of the final regulation. Firm recertification provides a mechanism for EPA to keep its records current with respect to firms actively engaged in renovations.

Recertification also provides a means for EPA to ensure that it has updated firm contact information. Recertification also prompts the firm to positively reaffirm its commitment to adhere to the requirements set forth in this regulation. Finally, recertification allows EPA an opportunity to review a firm's compliance history before it obtains recertification. However, EPA has determined that these purposes can be adequately served by recertifying renovation firms every 5 years instead of every 3 years as proposed.

▶ TRAINING PROVIDER ACCREDITATION AND RECORDKEEPING

EPA is amending the general accreditation requirements of 40 CFR 745.225 to apply to training programs that offer renovator or dust sampling technician courses for certification purposes. The regulations describe training program qualifications, quality control measures, recordkeeping and reporting requirements, as well as suspension, revocation, and modification procedures. Amendments to §745.225 add specific requirements for the renovator and dust sampling technician disciplines. Also included are minimum training curriculum, training hour, and hands-on requirements for courses leading to certification as a renovator or a dust sampling technician.

As discussed in the previous unit of the preamble, to assist EPA compliance inspectors in determining whether a renovator at a renovation work site successfully completed an accredited renovator training course, this final rule also requires providers of renovator training to take a digital photograph of each individual who successfully completes a renovator training course, include that photograph on the individual's course completion certificate, and provide that photograph to EPA along with the training course provider's post-training notification required by 40 CFR 745.225(c)(14).

Training course providers that obtained accreditation to offer renovator or dust sampling technician training would have to comply with the existing recordkeeping requirements for lead-based paint activities training course providers. These existing recordkeeping provisions require providers to maintain records of course materials, course test blueprints, information on how hands-on training is delivered, and the results of the students' skills assessments and course tests. EPA received no comments on this aspect of the proposed recordkeeping requirements. These requirements are currently working well for lead-based paint activities training providers and EPA believes they will work equally well for renovation training providers. Therefore, EPA is finalizing this requirement as proposed. Training course providers who receive accreditation to provide renovator or dust sampling technician courses must comply with the recordkeeping requirements of 40 CFR 745.225(i).

Renovator Training

The minimum curriculum requirements for an initial renovator course are described in 40 CFR 745.225(d)(6). The topics include the roles and responsibilities of a renovator; background information on lead and its health effects; background on applicable federal, state, and local regulations and guidance; use of acceptable test kits to test paint to determine whether it is lead-based paint; methods to minimize the creation of lead-based paint hazards during renovations; containment and cleanup methods; ways to verify that a renovation project has been properly completed, including cleaning verification; and waste handling and disposal. Hands-on activities relating to renovation methods, containment and cleanup, cleaning verification, and waste handling would be required in all courses. Section 745.225(c)(6)(vi) establishes the minimum length for an initial renovator course at 8 training hours, with 2 hours being devoted to hands-on activities.

Commenters raised concerns and had suggestions regarding how certified renovator training should be conducted in three broad areas: course length; course content and format; and training of non-English speaking renovators.

Course Length

Several commenters raised concerns about the length of the certified renovator training course. Some agreed with the training length as defined in the rule, others stated it was too short or too long, and one said that the length of the training should not be defined in the rule. In establishing the minimum requirements for the renovator course, the agency considered the many types of activities that would likely be performed during renovation, remodeling, and painting activities and tried to balance that with the need for a training course that would address the necessary skills without being overly burdensome on the part of the trainee.

The suggested course schedule for the EPA/HUD lead-safe work practices curriculum "Lead Safety for Remodeling, Repair, & Painting" calls for an 8-hour training day, including lunch, two breaks, and an hour-long course test. The course is designed in a modular format, so that it can be delivered in 1 day or over 2 or more days, at the discretion of the training provider. Based on a review of the material and the suggested schedule, EPA believes that "Lead Safety for Remodeling, Repair, and Painting" can be modified to include material on the use of test kits and performing cleaning verification and still fit within 8 training hours. However, any attempt to cover all of the required elements in a shorter period of time would likely result in a significant reduction in the level of detail with which the elements are presented. A minimum requirement for 8 training hours represents a reasonable minimum requirement for the renovator course and gives training course providers an indication of the amount of time that EPA has determined through experience with the EPA/HUD curriculum that it takes to adequately cover each required training element.

Course Content and Format

Most commenters agree that the certified renovator course should include a hands-on training portion and several of them agree that the hands-on portion should not be any shorter than 2 hours as proposed. Other commenters suggested that the hands-on portion of the training should be allowed to be conducted as a demonstration via a remote delivery system (DVD or Internet). EPA agrees that development of a procedure to address the hands-on component of the renovator course via remote delivery systems would be beneficial. This final rule does not preclude training providers from developing alternative methods for the delivery and evaluation of training for submission for approval to EPA.

Several commenters had suggestions as to the certified renovator training content. Two recommended that the renovator course include training on recordkeeping requirements. EPA agrees with these commenters, and has added the element of recordkeeping to the required training course elements for renovators. Because EPA has modified the recordkeeping requirements, as discussed below, to require the certified renovator to prepare the records associated with renovations to which he or she is assigned, the renovation training course will include a recordkeeping component. Three commenters suggested that, if the certified renovator is responsible for providing OJT to other renovation workers, the renovator training course should include a train-the-trainer component. EPA agrees with these commenters and has added a train-the-trainer element to the required elements for the renovator training course. In addition, EPA will develop a train-the-trainer component for its model renovator training course.

Other commenters suggested that the required training elements include OSHA health and personal safety requirements. The agency agrees that these are relevant topics and considers an overview of the OSHA requirements to be part of the required element of background on applicable federal, state, and local regulations and requirements. To ensure that this is clear, EPA has modified this provision to state that the background information must include EPA, HUD, OSHA, and other federal, state, and local regulations and guidance. Consistent with its approach in other courses related to lead-based paint activities, the agency believes that identifying potential OSHA requirements, rather than requiring in-depth curriculum components, is the best way to make trainees aware of those requirements and yet avoid redundancies between EPA- and OSHA-required courses.

Training of Non-English Speaking Renovators

Renovator and dust sampling technician courses, both initial and refresher, can be taught in any language, but accreditation would be required for each specific language the provider wished to present the course in. All course materials and instruction for the course would have to be in the language of the course. The modification to §745.225(b)(1)(ii) clarifies that all lead-based paint courses taught in different languages are considered different courses, and accreditation must be obtained for each.

To facilitate accreditation of courses in languages other than English, EPA is requiring that the training provider include in its application both the English version as well as the non-English version of all training materials, in addition to a signed statement from a qualified, independent translator that the translator has compared the non-English language version of the course materials to the English-language version and that the translation is accurate. This requirement applies to any course for which accreditation is sought, including lead-based paint activities courses. Finally, to assist EPA in monitoring compliance with these requirements, EPA is requiring that course completion certificates include the language in which the course was taught.

Several commenters agreed that the needs of non-English speaking workers should be considered. Commenters suggested that EPA translate its model course into other languages and/ or facilitate free access to such translations. EPA agrees that it is important to have renovator training available in languages other than English. EPA anticipates translating its revised model renovator course into Spanish. EPA will also consider translating the course into other languages. However, EPA is not able to make available proprietary material developed by training course providers that is then translated by those providers into other languages.

Dust Sampling Technician Training

The minimum curriculum requirements for an initial dust sampling technician course are described in 40 CFR 745.225(d)(7). The topics include the roles and responsibilities of a dust sampling technician; background information on lead and its adverse health effects; background information on federal, state, and local regulations and guidance that pertains to lead-based paint and renovation activities; dust sampling methodologies; clearance standards and testing; and report preparation and recordkeeping requirements. Section 745.225(c)(6)(vii) establishes the minimum length for an initial dust sampling technician course at 8 training hours, with 2 hours being devoted to hands-on activities.

EPA received relatively few comments specifically on the content of dust sampling technician training; most had to do with the length of the training course. EPA has developed a model dust sampling technician course. This course has been designed to be delivered in one 8-hour training day, including lunch, breaks, and a course test. As with the EPA/HUD "Lead Safety for Remodeling, Repair, & Painting" curriculum, EPA believes that this is a reasonable minimum requirement for the dust sampling technician course and it gives training course providers an indication of the amount of time that EPA has determined it takes to adequately cover each required training element.

The EPA Lead Rule contains extensive information. This is why so many people have become frustrated with it. This chapter has compiled key elements of training and certification. In the next chapter, we will look at work practices.

Work Practices

8

The EPA final rule requires that all renovations subject to this rule be conducted in accordance with a defined set of work practice standards. Again, this final rule is a revision of the existing TSCA section 402(a) Lead-Based Paint Activities Regulations to extend training, certification, and work practice requirements to certain renovation and remodeling projects in target housing and child-occupied facilities (COFs). In so doing, EPA did not merely modify the scope of the current abatement requirements to cover renovation and remodeling activities. Rather, EPA has carefully considered the elements of the existing abatement regulations and is revising those regulations in a manner that reflects the differences between abatement and renovation activities.

Work practices for abatement are part of a larger range of activities that are intended to identify and eliminate lead-based paint hazards. When abatements are conducted, residents typically are removed from the home until after the abatement activities are completed, which is demonstrated through the use of clearance testing. This may require the removal of carpeting, refinishing, sealing, or replacement of floors to achieve clearance. Accordingly, clearance testing is part of a broader set of activities that comprise abatement, with the purpose of permanently eliminating existing lead-based paint hazards.

Renovation, repair, and painting activities typically are conducted while the residents are present in the dwelling and are not activities intended to eliminate lead-based paint hazards. Work practices for renovation, repair, and painting are designed to minimize exposure to lead-based paint hazards created by the renovation both during the renovation, while residents are likely to be present in the dwelling, and after the renovation. The work practices are not intended to address preexisting hazards.

This final rule incorporates work practice standards generally derived from the HUD Guidelines, EPA's draft technical specifications for renovations, and the model training curriculum entitled "Lead Safety for Remodeling, Repair, and Painting."

To reduce exposure to lead-based paint hazards created by renovation activities, the work practices standards in this regulation provide basic requirements for occupant protection, site preparation, and cleanup.

Commenters generally felt that work practices are important and should be clear and correctly followed. One commenter stated that the rule has "tremendous potential for making a difference," especially in establishing and "reinforcing the industry norm." One commenter noted that EPA should "set simple and flexible work practices." Another commenter asked for less specificity. EPA believes that this final rule provides certified renovators an appropriate blend of flexibility and specificity. EPA believes that, due to the highly variable nature of renovation activities, flexibility is needed for certain tasks, such as establishing containment, and that other tasks, such as specialized cleaning, require a greater degree of specificity.

This final rule requires the firm to post signs clearly defining the work area and warning occupants and other persons not involved in renovation activities to remain outside of the work area. In addition, it requires that the certified renovator be physically present at the work site when the required signs are posted. These signs must be posted before beginning the renovation and must remain in place until the renovation has been completed and cleaning verification has been completed. The signs must be, to the extent practicable, provided in the occupants' primary language. If warning signs have been posted in accordance with HUD's Lead-Safe Housing Rule (24 CFR 35.1345(b)(2)) or OSHA's Lead in Construction standard (29 CFR 1926.62(m)), additional signs are not required.

Three commenters stated that the required signs for posting at a work site should be in the language of the occupant. One commenter stated that such a requirement would be consistent with HUD's Lead-Safe Housing Rule requirements. EPA agrees that having signs in the language of the occupant is preferable. However, the agency is concerned that renovators will not have the ability to provide signs in every language, and that it may be the case that occupants, especially in multifamily dwellings, will speak a variety of languages. In the HUD Lead-Safe Housing Rule, HUD addressed this issue by requiring that signs, to the extent practicable, be provided in the occupants' primary language. Therefore, consistent with HUD's Lead-Safe Housing

Rule, this final rule requires warning signs, to the extent practicable, to be provided in the occupants' primary language.

This final rule requires that the firm isolate the work area so that dust or debris does not leave the work area while the renovation is being performed. In addition, EPA has clarified that the firm must maintain the integrity of the containment by ensuring that any plastic or other impermeable materials are not torn or displaced, and taking any other steps necessary to ensure that dust or debris does not leave the work area while the renovation is being performed.

In addition, EPA has made conforming changes to the performance standard that renovators and renovation firms are being held to in this final rule. EPA was concerned that the rule text and preamble were confusing because there were references to "visible" dust and debris or "identifiable" dust and debris and "all" dust and debris. For example, in the 2006 Proposal "work area" was defined as the area established by the certified renovator to "contain all the dust and debris generated by a renovation." In the renovator responsibilities the renovator was responsible for ensuring "that dust and debris is not spread beyond the work area."

In describing the containment to be established, the rule text referred to "visible" dust and debris and in the section on waste from renovations the rule text referred to "identifiable" dust. It was not EPA's intention to create subjectivity as to whether dust and debris were being dispersed. By conforming its terminology EPA is clarifying that certified renovators and renovation firms must ensure that the dust and debris (as opposed to "visible" or "identifiable" dust and debris) generated by the renovation is contained. Should an EPA inspector observe dust or debris escaping from the containment, the certified renovator and the renovation firm would be in violation of this final rule.

This final rule also requires that the certified renovator be physically present at the work site when the required containment is established. This means the certified renovator must determine for each regulated project the size and type of containment necessary to prevent dust and debris from leaving the established work area. This determination will be based on the certified renovator's evaluation of the extent and nature of the activity and the specific work practices that will be used.

► CONTAINMENT

Containment refers to methods of preventing leaded dust from contaminating objects in the work area and from migrating beyond the work area. It includes, among other possible measures, the use of disposable plastic drop cloths to cover floors and objects in the work area, and sealing of openings with plastic sheeting where necessary to prevent dust and debris from leaving the work area. When planning a renovation project, it is the certified renovator's responsibility to determine the type of work site preparation necessary to prevent dust and debris from leaving the work area.

Renovation projects generate varying amounts of leaded dust, paint chips, and other lead-contaminated materials depending on the type of work, area affected, and work methods used. Because of this variability, the size of the area that must be isolated and the containment methods used will vary from project to project. Large renovation projects could involve one or more rooms and potentially encompass an entire home or building, while small projects may require only a relatively small amount of containment. The necessary work area preparations will depend on the size of the surface(s) being disturbed, the method used in disturbing the surface, and the building layout. For example, repairing a small area of damaged drywall would most likely require the containment of a smaller work area and less preparation than demolition work, which would most likely require a containment of a larger work area and more extensive preparation to prevent the migration of dust and debris from the work area.

The Environmental Field Sampling Study found that the following activities created dust-lead hazards at a distance of 6 feet from where the work was being performed:

- Paint removal by abrasive sanding
- Window replacement
- HVAC duct work
- Demolition of interior plaster walls
- Drilling into wood
- Sawing into wood
- Sawing into plaster

Based on these data, EPA believes that at least 6 feet of containment is necessary to contain dust generated by most renovation projects.

Under this final rule, at a minimum, interior work area preparations must include removing all objects in the work area or covering them with plastic sheeting or other impermeable material. This includes fixed objects, such as cabinets and countertops, and objects that may be difficult to move, such as appliances. Interior preparations must also include closing all forced air HVAC ducts in the work area and covering them with plastic sheeting or other impermeable material; closing all windows in the work area; closing and sealing all doors in the work area; and covering the floor surface in the work area, including installed carpet, with taped-down plastic sheeting or other impermeable material in the work area 6 feet beyond the perimeter of surfaces undergoing renovation or a sufficient distance to contain the dust, whichever is greater.

To ensure that dust and debris do not leave the work area, it may be necessary to close forced air HVAC ducts or windows near the work area. Doors within the work area that will be used while the job is being performed must be covered with plastic sheeting or other impermeable material in a manner that allows workers to pass through, while confining dust and debris to the work area. In addition, all personnel, tools, and other items, including the exterior of containers of waste, must be free of dust and debris when leaving the work area.

For exterior projects, the same performance standard applies; namely, the certified renovator or a worker under the direction of the certified renovator must contain the work area so that dust or debris does not leave the work area while the renovation is being performed. Additionally, in response to comments suggesting that EPA follow the HUD Guidelines with respect to exterior containment requirements, EPA has incorporated a similar 10-foot minimum. Consequently, this final rule requires that exterior containment include covering the ground 10 feet beyond the perimeter of surfaces undergoing renovation or a sufficient distance to collect falling paint debris, whichever is greater, unless the property line prevents 10 feet of such ground covering. EPA has concluded that this is an appropriate and reasonable precaution for exterior work, given the fact that some amount of dispersal of dust or debris is likely as a result of air movement, even on relatively calm days. In addition, EPA sees value in maintaining appropriate consistency between this regulation and related HUD rules and guidelines.

In addition to such ground covering, exterior work area preparations must include, at a minimum, closing all doors and windows within 20 feet of the outside of the work area on the same floor as the renovation, and closing all doors and windows on the floors below that area. For example, if the renovation involves sanding a 5 × 5-foot area of paint in the middle of the third floor of a building, and that side of the building is only 40 feet long, all doors and windows on that side of the third floor must be closed, as well as all of the doors and windows on that side of the second and first floors. In situations where other buildings are in close proximity to the work area, where the work area abuts a property line, or weather conditions dictate the need for additional containment (i.e., windy conditions) the certified renovator or a worker under the direction of the certified renovator performing the renovation may have to take extra precautions in containing the work area to ensure that dust and debris from the renovation does not contaminate other buildings or migrate to adjacent property.

This may include erecting vertical containment designed to prevent dust and debris from contaminating the ground or any object beyond the work area. In addition, doors within the work area that will be used while the job is being performed must be covered with plastic sheeting or other impermeable material in a manner that allows workers to pass through while confining dust and debris to the work area.

Some commenters agreed with the proposed procedures. One commenter agreed that with containment, dust can be contained and cleaned up sufficiently to pass the wipe test screening results. Another commenter supported the use of standard containment and cleaning practices known to reduce dust lead levels on both interior and exterior surfaces and to protect soils and gardens surrounding the house.

Some commenters asserted that the containment procedures were not stringent enough. Some suggested that EPA follow the HUD Guidelines with respect to exterior containment requirements. Others asked EPA to strengthen exterior containment requirements by specifying that containment extend at least 20 feet to collect all debris and residue and that the rule address circumstances such as wind and rain. One commenter asserted that allowing the certified renovator complete discretion to determine what is appropriate renders the worksite containment requirements completely unenforceable and asked EPA to consider providing a minimum performance standard that all renovators must meet. EPA agrees that a minimum performance standard is necessary and that is why under this final rule EPA is requiring certified renovators to establish containment that prevents dust and debris from leaving the work area.

In addition, in this rule EPA has established minimum containment requirements for both interior and exterior renovation requirements. While the certified renovator has discretion regarding the specific components and extent of containment, the renovator and firm will be in violation of this final rule if dust or debris leaves the work area for both interior and exterior renovations. If dust or debris migrates beyond the work area, that migration constitutes a violation of the rule. Accordingly, EPA does not agree with the commenter that the rule is unenforceable.

This final rule provides the certified renovator with some discretion to define the specific size and configuration of the containment to accommodate the variability in size and scope of renovations. EPA considered requiring that in all cases the entire room in which a renovation is occurring be contained, but concluded that doing so would be unwarranted. For example, a small manual sanding job in a large room would not necessarily require full room containment to isolate the work area. EPA has concluded that the most appropriate approach is to impose a minimum size for containment coupled with a performance standard—preventing dust or debris from leaving the work area—and to prescribe with reasonable specificity the containment measures that are required (e.g., use of plastic of other impermeable material, removal or covering of objects in the work area) but to provide some measure of discretion with regards to the case-specific approaches to containment.

In response to EPA's request for comments on whether there are any situations where some or all of the proposed work practices are not necessary, commenters suggested that work practices were not needed during a gut rehabilitation, although two of the commenters suggested a waiver rather than an exemption in these situations. Several commenters thought that work in unoccupied structures should not require the use of lead-safe work practices, or should have an adapted set of work practices.

A commenter opined that certain interior containments may not be necessary in vacant and empty housing, but that exterior work always should use lead-safe work practices to protect the environment and neighborhood. A commenter stated that there are certain activities common to multifamily and rental housing that warrant special consideration from the agency. For example, simple painting activities that occur when rental properties turn over should not require a full suite of work practices, particularly given that most state laws require apartment owners to paint each unit at turnover. The commenter suggested that EPA consider a less restrictive set of guidelines for those properties simply undergoing routine painting during the turnover process.

EPA believes that whole house gut rehabilitation projects may demolish and rebuild a structure to a point where it is effectively new construction. In this case, it would not be a modification of an existing structure, and therefore not a renovation. However, a partial-house gut rehabilitation such as a kitchen or bathroom gut rehabilitation project clearly falls within the scope of this final rule.

EPA disagrees that temporarily unoccupied or vacant housing should be *per se* exempt from the requirements of this final rule. EPA's primary concern with exempting renovations in such housing from the work practices required by this final rule is the exposure to returning residents to lead-based paint hazards created by the renovation. However, EPA recognizes that if no child under 6 or no pregnant woman resides there, the owner-occupant may so state in writing and the requirements of this rule would not apply.

In addition, for routine painting, such as at unit turnover, if such painting activity does not involve disturbing more than 6 ft^2 of painted surfaces per room for interiors or 20 ft^2 for exteriors, and otherwise meets the definition of "minor repair and maintenance," the requirements of this final rule would not apply. EPA cannot see a basis for imposing a less restrictive set of requirements for projects that disturb more than 6 ft^2 of painted surfaces per room for interiors or 20 ft^2 for exteriors.

Some commenters believed that the proposal did not adequately address the decontamination of workers and equipment involved in a renovation. They supported the proposed requirement that all personnel, tools, and other items, including the exteriors of containers of waste, be free of dust and debris before leaving the work area.

However, they believed that the proposed alternative, covering the paths used to reach the exterior of the building with plastic, was not sufficiently protective. One contended that significant lead dust contamination can be tracked or carried out of a work area if workers and equipment are not properly decontaminated. This commenter further noted that workers with contaminated clothing can take that contamination home to their own children and taking contaminated equipment to another jobsite could potentially create a lead hazard at a new site. EPA agrees with these commenters and has deleted the alternative language.

The final rule requires renovation firms to use precautions to ensure that all personnel, tools, and other items, including the exteriors of containers of waste, be free of dust and debris before leaving the work area. There are several ways of accomplishing this. For example, tacky mats may be put down immediately adjacent to the plastic sheeting covering the work area floor to remove dust and debris from the bottom of the workers' shoes as they leave the work area, workers may remove their shoe covers (booties) as they leave the work area, and clothing and materials may be wet-wiped and/or HEPA-vacuumed before they are removed from the work area.

Finally, in response to a commenter who was concerned about containment not impeding occupant egress in an emergency, EPA has modified the regulatory text to specify that containment must be installed in such a manner that it does not interfere with occupant and worker egress in an emergency. This can be accomplished, as noted in Chapter 17 of the HUD Guidelines, by installing plastic over doors with a weak tape.

► PROHIBITED AND RESTRICTED PRACTICES

The final rule prohibits or restricts the use of certain work practices during regulated renovations. These practices are open-flame burning or torching of lead-based paint; the use of machines that remove lead-based paint through high-speed operation such as sanding, grinding, power planing, needle gun, abrasive blasting, or sandblasting, unless such machines are used with HEPA exhaust control; and operating a heat gun above 1100° Fahrenheit. These are essentially the same practices as are currently prohibited or restricted under the Lead-Based Paint Activities Regulations, 40 CFR 745.227(e)(6), with the exception of dry hand scraping of lead-based paint. While this final rule and EPA's regulations do not prohibit or restrict the use of volatile paint strippers or other hazardous substances to remove paint, the use of these substances is prohibited in poorly ventilated areas by HUD's Lead-Safe Housing Rule and they are regulated by OSHA.

EPA did not propose to prohibit or restrict any work practices, but instead asked for public comment regarding their prohibition or restriction. The agency was concerned that, because these practices are commonly used during renovation work, prohibiting such practices could make certain jobs, such as preparing detailed or historic millwork for new painting, extremely difficult, if not impossible. In addition, EPA believed that use of the proposed package of training, containment, cleanup, and cleaning verification requirements would be effective in preventing the introduction of new lead-based paint hazards, even when such practices were used. EPA is modifying the proposal based on new data evaluating specific work practices and in response to comments received.

EPA's Dust Study
EPA understood when developing the proposed rule that considerable data existed showing the potential for significant lead contamination when lead paint is disturbed by practices restricted under EPA's Lead-Based Paint Activities Regulations for abatements. EPA conducted the Dust Study, in part, to determine the effectiveness of the proposed work practices. The Dust Study evaluated a variety of renovation activities, including activities that involved several practices restricted or prohibited under the abatement regulations. For example, power planing was included in the Dust Study as a representative of machines that remove lead-based paint through high speed operation. Similarly, the Dust Study also included experiments with power sanding and a needle gun. Each of these activities generated very high levels of dust. The Dust Study thus evaluated the proposed work practice standards, using a range of typical practices currently used by contractors.

In particular, the Dust Study found that renovation activities involving power planing and high temperature heat gun resulted in higher post-job renovation dust lead levels than activities using other practices. The geometric mean post-work, precleaning floor dust lead levels in the work room were 32,644 µg/ft^2 for power planing and 7737 µg/ft^2 for high temperature heat guns. More importantly, in experiments performed in compliance with this rule's requirements for containment, cleaning, and cleaning verification, the geometric mean post-job floor dust lead levels were still 148 µg/ft^2 for power planing, well over the TSCA section 403 hazard standard for floors.

While the geometric mean post-job floor dust levels for the three similar experiments involving high temperature heat guns (i.e., experiments performed in compliance with this rule's requirements) were 36 µg/ft^2, the average post-cleaning-verification floor dust lead levels for the individual experiments were 147.5, 65.5, and less than 10 µg/ft^2. Thus, in two of these three experiments, the requirements of this final rule were insufficient to reduce the floor dust lead levels below the TSCA section 403 hazard standards for floors. In addition, power planing and use of a high temperature heat gun generated fine particle-size dust that was difficult to clean. In fact, almost all of the high post-renovation lead levels were associated with activities involving power planing and high temperature heat guns. Moreover, activities involving power planing and high temperature heat gun jobs also resulted in higher post-job tool room and observation room lead levels than other practices.

Thus, while the Dust Study confirmed that most practices prohibited or restricted under EPA's Lead-Based Paint Activities Regulations do indeed produce large quantities of lead dust, it also demonstrated that, with respect to lead-based paint hazards created by machines that remove lead-based paint through high-speed operation and high temperature heat guns, the use of the proposed work practices were not effective at containing or removing dust-lead hazards from the work area.

Alternatives to Certain Practices

As discussed above, in the proposed rule, EPA stated a concern that, because practices prohibited or restricted under EPA's Lead-Based Paint Activities Regulations are commonly used during renovation work, prohibiting or restricting such practices could make certain jobs, such as preparing detailed or historic millwork for new painting, extremely difficult or, in some cases, impossible. In response to its request for comment, the agency received information on techniques including benign strippers, steam stripping, closed planing with vacuums, and infrared removal that the commenter believed are far superior, far safer, and far cheaper than some of the traditionally prohibited or restricted practices.

Another commenter noted that window removal and offsite chemical stripping in a well-ventilated setting is an alternative to using heat or mechanical methods to remove lead paint onsite. Alternatively, chemical strippers can be used onsite, given adequate ventilation and protection for workers and building occupants. EPA is therefore persuaded that there are sufficient alternatives to these practices.

Based on the results of the Dust Study and in response to the voluminous persuasive public comments, this final rule prohibits or restricts the use of the following practices during renovation, repair, and painting activities that are subject to the work practice requirements of this rule:

- Open-flame burning or torching
- Machines that remove lead-based paint through high-speed operation such as sanding, grinding, power planing, needle gun, abrasive blasting, or sandblasting, unless such machines are used with HEPA exhaust control
- Operating a heat gun above 1100°F

EPA has concluded that these practices must be prohibited or restricted during renovation, repair, and painting activities that disturb lead-based paint because the work practices in this final rule are not effective at containing the spread of leaded dust when these practices are used, or at cleaning up lead-based paint hazards created by these practices. Thus, the work practices are not effective at minimizing exposure to lead-based paint hazards created during renovation activities when these activities are used.

There are many interim controls that eliminate the need for disturbing lead-based paint. These methods are proven and should be considered. For example, new trim can be installed over hazardous surfaces. Installing underlayment and new flooring over a contaminated floor will contain the risk without requiring removal of the dangerous materials. For examples of this type of interim control, see Figures 8.1, 8.2, 8.3, and 8.4. Box 8.1 contains a list of techniques for removing loose paint on impact surfaces using sticky tape.

Figure 8.1 Installing underlayment over older flooring as a suitable lead-based paint enclosure method.

This final rule does not prohibit or restrict the use of dry hand scraping. EPA has concluded based primarily on the Dust Study as corroborated by other data described below that it is not necessary to prohibit or restrict dry scraping because the containment, cleaning, and cleaning verification requirements of this rule are effective at minimizing exposure to lead-based paint hazards created by renovations and the migration of dust-lead hazards beyond the work area when dry hand scraping is employed.

The Dust Study evaluated dry hand scraping, which is restricted under EPA's lead abatement program. In contrast to the results of the activities using power planing and high temperature heat gun, average post-job dust lead levels in the two experiments in which paint was disturbed by dry hand scraping

Remove the bottom sash. If the counter-weight ropes or chains are in place, do not let them drop into the weight compartment.

Remove the paint from edges that rub against stop, stool, and parting bead. Wet planing is a good method.

Rehang the sash(es) in a compression track. If there is no counter weight or spring system, install one to keep the sash in place.

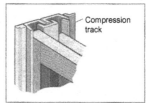

Figure 8.2 Reducing friction with window channel guides.

Enclose risers with thin plywood (like luan plywood) or some other hard material. Whatever you use must fit snugly.

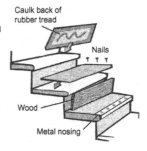

Back caulk the edges of treads. Place them and nail or screw them down. Screw or nail the metal nosing on.

Figure 8.3 Covering stair treads with tread guards.

Note: A rubber tread with metal nosing works well. Rubber nosing that fits snugly on the nose may work if the stairs are not used very often.

and the work practices required by this rule were used were below the regulatory dust-lead hazard standard for floors. In addition, the National Institute for Occupational Safety and Health (NIOSH) conducted a Health Hazard Evaluation (HHE) at the request of the Rhode Island Department of Health, and published a final report in June 2000.

The purpose of the evaluation was to measure worker exposure during various tasks and to determine whether workers were exposed to hazardous amounts of lead-based paint.

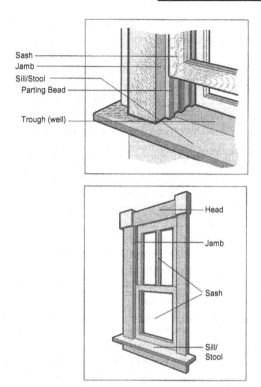

Figure 8.4 Window friction and impact points.

Labels (top diagram): Sash, Jamb, Sill/Stool, Parting Bead, Trough (well)

Labels (bottom diagram): Head, Jamb, Sash, Sill/Stool

Box 8.1   Using Sticky Tape for Abatement

1. Place a piece of plastic or paper beneath the area in question.
2. Press a piece of wide sticky tape firmly over the area of loose/chipping paint.
3. Wait a few seconds and then carefully remove the tape, taking the small chips of paint with it.
4. Place the tape in a plastic bag.
5. Carefully fold the piece of plastic or paper that was beneath the area and place it in the bag.
6. Seal the bag and clean the area.
7. Dispose of all waste materials in a secure manner; do not use the resident's trash cans for this purpose.

Notably, worker exposures were compared when scraping paint-ed surfaces using wet and dry scraping methods (wet scraping is the customary substitute for dry scraping in abatement applications). A comparison of worker exposure found statistically equivalent worker exposures. Based on the NIOSH study, EPA has determined that dry scraping is the equivalent of its only practical alternative, wet scraping.

In sum, EPA has determined based on the studies described above and the persuasive comments, including those summarized below, provided by the overwhelming majority of commenters that its approach of prohibiting or restricting certain practices in combination with the containment, cleaning, and cleaning verification, will be effective in minimizing exposure to lead-based paint hazards created during renovation activities, provide an appropriate measure of consistency with other regulatory programs, and cause minimal disruption for renovation firms.

Numerous commenters argued that the rule should prohibit certain practices based on potential health hazards, many backed up by well-documented scientific studies and proven health-protective standards. One commenter stated, after citing several scientific studies, that removing or disturbing lead paint without proper controls causes substantial contamination, posing serious risks to occupants, workers, and others. Another cited numerous scientific studies demonstrating the adverse public health implications of permitting these work practices and the availability of alternative work methods. Still another cited the EPA renovation and remodeling study and a State of Maryland study as evidence that prohibited work practices may be associated with elevated blood-lead levels. One commenter cited health hazard evaluations of residential lead renovation work showing that these activities produce hazardous worker exposures.

Another commenter noted that the hazards of activities that are likely to produce large amounts of lead dust or fumes are well documented, stating that, for example, the Wisconsin Childhood Blood-Lead Study found that the odds of a resident child having a blood-lead level in excess of 10 μg/dL increased by 5 times after renovation using open flame torching, and by 4.6 times after heat gun use. Another commenter was concerned that previously collected data may not account for different particle-size distribution, a factor in both the potential cleaning efficacy of work areas and the toxicology of lead poisoning.

Some commenters urged EPA to prohibit certain high dust generating practices for the sake of consistency with other work practice standards. Numerous commenters asserted EPA's rule should be consistent with HUD requirements to avoid confusion on the part of contractors and to conform to the standard that has been in place for nearly 6 years. One commenter noted that the regulations of several other federal agencies that administer housing programs, such as the Department of Defense, Department of Agriculture, and Department of Veterans Affairs include prohibited practices. Other commenters noted that the proposed rule conflicted with OSHA rules and would cause confusion among contractors.

Some commenters noted that EPA's proposed rule would conflict with individual state or local regulations prohibiting some or all of these practices. One commenter listed the following states and some cities that have prohibited work practices: California, Indiana, Maine, Massachusetts, Minnesota, New Jersey, Ohio, Rhode Island, Vermont, Wisconsin, Chicago, Cleveland, New Orleans, New York City, Rochester, and San Francisco. Two commenters cited state law in Indiana, under which certain work practices are prohibited and contractors using such work practices are committing a Class D felony.

Other commenters noted that practices that are prohibited under EPA's Lead-Based Paint Activities Regulations should also be prohibited for renovation work in pre-1978 properties, and noted that in developing the abatement rule EPA demonstrated through its own studies that these practices may increase the risk of elevated blood-lead levels in children.

▶ WASTE FROM RENOVATIONS

Under this final rule the certified renovator or a worker trained by and under the direction of the certified renovator is required to ensure that all personnel, tools, and other items including waste are free of dust and debris when leaving the work area. The certified renovator or a worker trained by and under the direction of the certified renovator must also contain waste to prevent releases of dust and debris before the waste is removed from the work area for storage or disposal. If a chute is used to remove waste from the work area, it must be covered. At the conclusion of each work day and at the conclusion of the renovation, the certified renovator or a worker trained by and under the direction of the certified renovator must ensure that waste that has been collected from renovation activities is stored under containment, in an enclosure, or behind a barrier that prevents release of dust and debris from the work area and prevents access to dust and debris.

This final rule also requires the certified renovator or a worker trained by and under the direction of the certified renovator transporting lead-based paint waste from a work site to contain the waste to prevent releases, for example, inside a plastic garbage bag. As described in more detail in Unit IV.D.2.c. of the preamble to the 2006 Proposal, EPA revised its solid waste regulations in 40 CFR, Parts 257 and 258, to make clear that lead-based paint waste generated through renovation and remodeling activities in residential settings may be disposed of in municipal solid waste landfill units or in construction and demolition (C&D) landfills. Requirements for waste disposal may vary by jurisdiction and state and local requirements may be more stringent than federal requirements. When disposing of waste, including waste water, from renovation activities, the renovation firm must ensure that it complies with all applicable federal, state, and local requirements.

One commenter suggested that EPA should consider requiring that lead-contaminated waste be stored in a locked area or in a lockable storage container. This commenter also suggested that to prevent any confusion on what constitutes a covered chute, a definition or clarification should be provided in the rule. Another commenter recommended the use of "sealed" rather than "covered" chutes for waste removal, as a covered chute may not be protective enough to prevent the release of significant amounts of lead-contaminated dust. This final rule requires that waste must be contained to prevent releases of dust and debris before the waste is removed from the work area for storage or disposal.

With respect to the use of chutes for waste removal, the requirement for a covered chute was proposed merely to facilitate the removal of bagged or sealed waste so that it is deposited in an appropriate waste disposal container and does not fall to the ground. EPA does not, therefore, believe that this term either needs to be further defined or to require the use of a "sealed" chute.

EPA understands that renovation projects can generate a considerable amount and variety of waste material. However, EPA believes that the requirements of the final rule protect occupants and others from potential lead-based paint hazards presented by this waste. While storing the waste in a locked container is one way to meet the performance standard of this final rule, EPA does not believe that it is necessary to specify that as a requirement. The waste may be stored in the work area, which will already be delineated with signs cautioning occupants and others to keep out. EPA believes the owner/occupants have some responsibility for observing these signs. Renovation sites pose potential hazards other than lead-based paint hazards, including the potential for falls, sharp protrusions, and so on.

In sum, the certified renovator is responsible for ensuring that lead-contaminated building components and work area debris are stored under containment, in an enclosure, or behind a barrier that prevents release of dust and debris and prevents access to the debris. Under this final rule the certified renovator must ensure that waste leaving the work area is contained (e.g., in a heavy-duty plastic bag or sealed in plastic sheeting) and free of dust or debris. This imposes a reasonable performance standard without requiring a specific approach. The certified renovator is responsible for evaluating the waste generated and the characteristics of the work site to determine the most effective way of meeting this standard.

▶ CLEANING THE WORK AREA

Under this final rule the certified renovator or a worker under the direction of the certified renovator must clean the work area to remove dust, debris, or residue. All renovation activities that disturb painted surfaces can produce dangerous quantities of leaded dust. Because very small particles of leaded dust are easily absorbed by the body when ingested or inhaled, it can create a health hazard for children. Unless this dust is properly removed, renovation and remodeling activities are likely to introduce new lead-based paint hazards. Therefore, the rule requires prescriptive cleaning practices. Ultimately, improper cleaning can increase the cost of a project because additional cleaning may be necessary during post-renovation cleaning verification.

This final rule requires that, after completion of interior renovation activities, all paint chips and debris must be picked up. Protective sheeting must be misted and folded dirty side inward. Sheeting used to isolate contaminated rooms from noncontaminated rooms must remain in place until after the cleaning and removal of other sheeting; this sheeting must then be misted and removed last.

Removed sheeting must be either folded and taped shut to seal or sealed in heavy-duty bags and disposed of as waste. After the sheeting has been removed from the work area, the entire area must be cleaned including the adjacent surfaces that are within 2 feet of the work area. The walls, starting from the ceiling and working down to the floor, must be vacuumed with a HEPA vacuum or wiped with a damp cloth. The final rule requires that all remaining surfaces and objects in the work area, including floors, furniture, and fixtures be thoroughly vacuumed with a HEPA vacuum.

When cleaning carpets, the HEPA vacuum must be equipped with a beater bar to aid in dislodging and collecting deep dust and lead from carpets. The beater bar must be used on all passes on the carpet face during dry vacuuming. This cleaning step is intended to remove as much dust and remaining debris as possible. After vacuuming, all surfaces and objects in the work area, except for walls and carpeted or upholstered surfaces, must be wiped with a damp cloth. Wet disposable cleaning cloths of any color may be used for this purpose. In contrast, as discussed in the next section, only wet disposable cleaning cloths that are white may be used for cleaning verification. Uncarpeted floors must be thoroughly mopped using a two-bucket mopping method that keeps the wash water separate from the rinse water, or using a wet-mopping system with disposable absorbent cleaning pads and a built-in mechanism for distributing or spraying cleaning solution from a reservoir onto a floor.

When cleaning following an exterior renovation, all paint chips and debris must be picked up. Protective sheeting used for containment must be misted with water. All sheeting must be folded from the corners or ends to the middle to trap any remaining dust and either taped shut to seal or sealed in heavy-duty bags. The sheeting must be disposed of as waste.

Several commenters proposed minor changes to the cleaning procedures. Three commenters recommended that daily cleanup be required for projects lasting more than 1 day. One commenter stated that all tools and equipment should be cleaned prior to leaving the job site. One commenter indicated concern that there is no mention of wet wiping areas such as windowsills. This final rule requires cleaning both in and around the work area to ensure no dust or debris remains following the renovation. The final rule also requires that all personnel, tools, and other items including waste are free of dust and debris when leaving the work area. EPA recommends that contractors keep work areas as clean and free of dust and debris as practical. Daily cleaning is a good practice, and it may be necessary in some cases to ensure no dust or debris leaves the work area as required by this final rule. However, EPA has no basis to believe that daily cleaning is necessary in every case or even most cases. EPA also notes that the work area must be delineated by signs so that occupants and others do not enter the area.

This final rule requires the work area to be contained, and to ensure that all tools, personnel, and other items, including waste, be free of dust and debris when leaving the work area. Under this final rule, interior windowsills and most other interior surfaces in the work area must be wet wiped. The exceptions are upholstery and carpeting, which must be vacuumed with a HEPA vacuum, and walls, which may be wet wiped or vacuumed with a HEPA vacuum.

Some commenters requested clarification of the requirement to clean "in and around the work area." In response to the two commenters who noted that the HUD Guidelines recommend cleaning 2 feet beyond the work area, EPA has modified the regulatory text to require cleaning of surfaces and objects in and within 2 feet of the work area.

One commenter argued that vacuuming was not necessary because 40 CFR, Part 745.85, requires the certified renovator to cover all furnishings not removed from the work area, so additional cleaning is unnecessary. EPA disagrees with this commenter. Carpets and upholstered objects that remained, covered with plastic, in the work area during the renovation must be vacuumed after the plastic is removed to ensure that the surfaces did not become contaminated during the renovation due to a breach in the containment or during the removal of the containment during cleanup.

One commenter asserted that some requirements for cleaning were not prescriptive enough. The commenter suggested that the rule text, which states that the certified renovator or a worker under the direction of the certified renovator must "pick up all paint chips and debris," could be reworded to state that the certified renovator or a worker under the direction of the certified renovator must "collect all paint chips, debris, and dust, and, without dispersing any of it, seal this material in a heavy-duty plastic bag." EPA agrees that additional detail would be helpful in this instance and has modified the final rule to include this recommendation, with the exception of dust, which is collected when the protective sheeting is misted and folded inward.

One commenter stated that the cleaning procedures were excessive and problematic. This commenter asserted that the two-bucket mopping system is inappropriate for some floor types such as wood floors for which excessive water could damage the floor. The commenter suggested that EPA allow a cleaning method employing a dry or damp cloth, or any other specified methodology, to be used to achieve a no-dust or no-debris level of cleaning. Three commenters asserted that EPA's definition of wet-mopping system was too specific. One commenter stated that to include "a long handle, a mop head . . ." in the description of the wet-mopping system is too prescriptive and favors a particular model of commercial product. EPA understands that the two-bucket mopping system may not be appropriate for all floor types due to the quantity of water involved. However, the HUD Guidelines recommend and the Dust Study demonstrates that wet cleaning is best able to achieve desired results. This final rule allows for the use of a wet-mopping system instead of the two-bucket system for the cleaning of flooring. EPA has included a definition of a wet-mopping system to allow the regulated community to use such a system in place of the traditional two-bucket mop method.

Electrostatic Cloth and Wet Cloth Field Study in Residential Housing

A study indicates that a wet-mopping system is an effective method for cleaning up leaded dust. EPA believes that allowing the use of a wet-mopping system like those widely available in a variety of stores should alleviate concerns regarding the quantity of water used in the cleanup. In addition, EPA disagrees that the description of a wet-mopping system favors a particular model of commercial product. Rather, it generally describes any number of wet-mopping systems widely available in most stores. However, to alleviate concerns that a particular model of commercial product is preferred, EPA has added the phrase "or a method of equivalent efficacy" to the end of the definition of "wet-mopping system."

One commenter recommended that instead of referencing a two-bucket method, EPA should consider simply stating that a method be used that keeps the wash water separate from the rinse water. EPA agrees and has revised the regulatory text to specify a method that keeps wash water separate from rinse water, giving as an example the two-bucket method.

One commenter questioned the requirement to vacuum underneath a rug or carpet where feasible. The commenter suggested that EPA clarify that this does not include permanently affixed wall-to-wall carpeting. The commenter notes that it is highly unlikely that the renovation or remodeling activity conducted in a carpeted room would have created the dust embedded underneath both the layer of plastic sheeting and the installed carpeting. EPA agrees with this commenter. EPA did not intend to require vacuuming beneath permanently affixed carpets (i.e., wall-to-wall carpeting), but rather that removable rugs should be removed and the area beneath vacuumed. However, small, movable, area rugs should be removed from the work area prior to the renovation and the floor beneath would be cleaned as required under this final rule. Therefore, in response to this commenter, EPA has deleted the requirement to vacuum beneath rugs where feasible.

One commenter recommended four options for cleaning carpets: Removing the carpet and pad, cleaning the underlying flooring, then replacing the carpet and pad; shampooing the carpet using a vacuum attachment that removes the suds; steam cleaning the carpet using a vacuum attachment that removes the moisture; or HEPA-filtered vacuuming. This final rule seeks to minimize the introduction of lead-based paint hazards to carpeted floors by requiring the certified renovator to cover the floor of the work area with plastic sheeting, carefully clean up and remove the plastic sheeting following work, and thoroughly vacuum the carpet using a HEPA vacuum with a beater bar. EPA believes this containment and cleanup protocol will minimize exposure to lead-based paint hazards created during renovation activities.

EPA does not believe a renovation contractor should be responsible for removing and replacing carpet in a home when such a requirement was not within the scope of the renovation project. Also, in contrast to the effectiveness of using a HEPA on carpets, EPA does not have sufficient data on steam cleaning or shampooing to evaluate its effectiveness. Without data to demonstrate the effectiveness of shampooing or steam cleaning carpets EPA is not prepared to require these methods be used in lieu of vacuuming with a HEPA vacuum. EPA further notes that the HUD Lead-Safe Housing Rule only requires HEPA vacuuming, not steam cleaning or shampooing.

▶ VACUUMS EQUIPPED WITH HEPA FILTERS

Given that the HUD guidelines recommend the use of HEPA vacuums and the OSHA Lead in Construction standard requires that vacuums be equipped with HEPA filters where vacuums are used, EPA proposed requiring the use of HEPA vacuums in its proposed work practices. Nonetheless, EPA requested comment on whether the rule should allow the use of vacuums other than vacuums equipped with HEPA filters. Specifically, EPA requested comment on whether there are other vacuums that have the same efficiency at capturing the smaller lead particles as HEPA-equipped vacuums, along with any data that would support this performance equivalency and whether this performance specification is appropriate for leaded dust cleanup.

HEPA filters were first developed by the U.S. Atomic Energy Commission during World War II to capture microscopic radioactive particles that existing filters could not remove. HEPA filters have the ability to capture particles of 0.3 microns with 99.97% efficiency. Particles both larger and smaller than 0.3 microns are easier to catch. Thus, HEPA filters capture those particles at 100%. Available information indicates that lead particles generated by renovation activities range in size from over 20 microns to 0.3 microns or less.

OSHA recently completed a public review of its Lead in Construction standard. OSHA concluded that the principal concerns regarding HEPA vacuums (i.e., cost and availability) have been significantly reduced since the standard was established in 1994. HEPA vacuum cleaners have an increased presence in the marketplace and their cost has decreased significantly. Therefore,

OSHA continues to require the use of HEPA vacuums in work subject to the Lead in Construction standard.

Vacuums used as part of the work practices being finalized in this final rule must be HEPA vacuums, which are to be used and emptied in a manner that minimizes the reentry of lead into the workplace. The term "HEPA vacuum" is defined as a vacuum that has been designed with a HEPA filter as the last filtration stage. A HEPA filter is a filter that is capable of capturing particles of 0.3 microns with 99.97% efficiency. The vacuum cleaner must be designed so that all the air drawn into the machine is expelled through the filter with none of the air leaking past it. Many commenters supported the use of HEPA vacuums. Some of these commenters supported the requirement that they be used because they are also required by the OSHA Lead in Construction standard. One commenter noted that the price of HEPA vacuums had decreased and they were no longer significantly more expensive than non-HEPA vacuums.

Another commenter cited the Dust Study, the NAHB Lead-Safe Work Practices Survey, and several other studies as supporting the conclusion that lead-safe work practices and modified lead-safe work practices, along with a two-step or three-step cleaning process using a HEPA-equipped vacuum and wet washing, greatly reduce dust lead levels and should be regarded as best management practices for renovation jobs. The commenter notes that the NAHB study found significant reductions in loading levels after cleanup using a HEPA-equipped vacuum and then either wet washing or using a wet disposable cleaning cloth mop.

One commenter contended that HEPA vacuums with beater bars were not currently available on the market at the time comments were submitted. However, EPA has been able to identify commercial vacuum manufacturers as well as department store brands that currently offer HEPA vacuums with beater bar attachments.

Several commenters noted that vacuum cleaners other than HEPA vacuums were effective at removing lead dust. They cited several papers that they asserted support their conclusion. They include Comparison of Home Lead Dust Reduction Techniques on Hard Surfaces: The New Jersey Assessment of Cleaning Techniques Trial by Rich et al. (*Environmental Health Perspective*, 110(9):889-93, 2002), a study by the California

Department of Health Services that the commenter contends concluded that some non-HEPA vacuums performed better than the HEPA units tested; Comparison of Techniques to Reduce Residential Lead Dust on Carpet and Upholstery: The New Jersey Assessment of Cleaning Techniques Trial by Yiin et al. (*Environmental Health Perspective*, 110(12):1233-37, 2002); and *Effectiveness of Clean-up Techniques for Leaded Paint Dust* by Makohon et al. (Canada Mortgage and Housing Corporation, Research Division, 1992).

The commenter who cited the Rich et al. paper contended that the authors found no clear difference between the efficacy of HEPA and non-HEPA vacuums on hard surfaces (noncarpeted floors, windowsills, and window troughs), and found that non-HEPA vacuums appeared more efficient in removing particles on uncarpeted floors, which are the hard surfaces that may best reflect exposure to children. One commenter stated that given the research literature demonstrates that there is no performance difference in lead dust removal, EPA should allow cleanup with either a HEPA or non-HEPA vacuum. Another commenter contended that a vacuum cleaner retrofitted with a HEPA filter rather than a HEPA vacuum should be required to be used as part of the work practices.

EPA disagrees with the commenters who state that the literature does not demonstrate a difference between HEPA vacuums and non-HEPA vacuums. In the Yiin et al. study, the authors stated that for carpets, data from the "Environmental and Occupational Health Sciences Institute vacuum sampling method showed a significant reduction (50.6%, $p = 0.014$) in mean lead loading for cleaning using the HEPA vacuum cleaner but did not result in a significant difference (14.0% reduction) for cleaning using the non-HEPA vacuum cleaner." They also note that when they used wipe sampling "the results indicated that neither of the cleaning methods yielded a significant reduction in lead loading."

EPA believes the results from the wipe sampling method are less useful because as discussed in Unit III.E.8.iv. of this preamble, the agency believes that wipe sampling on carpets is not a reliable indicator of the lead-based paint dust in the carpet. The authors report that in their study non-HEPA vacuums were more effective than HEPA vacuums on upholstery but note "The reduced efficiency of the HEPA vacuum cleaner in cleaning upholstery, as compared to carpets, may be, at least partially, due to the lower precleaning dust lead level and the smaller sample data set for the HEPA vacuum cleaner than for the non-HEPA vacuum cleaner."

In the Rich et al. study, the authors noted: "On windowsills, the HEPA vacuum cleaner produced 22% (95% CI, 11–32%) larger reductions than the non-HEPA vacuum cleaner, and on the window troughs it produced 16% (95% CI, 4–33%) larger reductions than the non-HEPA vacuum cleaner." Not only were the percent reductions greater, the post-cleaning geometric mean lead loadings for the experiments in which the HEPA vacuums were used were lower than the post-geometric mean lead loadings for the experiments in which the non-HEPA vacuums were used.

On hard floors, the authors reported that the non-HEPA vacuum removed the largest quantities of lead-based paint dust. They note that this may be due in part to the fact that the initial loadings were higher where the non-HEPA vacuums were used (precleaning geometric mean lead loadings were 200 and 155 $\mu g/ft^2$ for the two types of experiments where non-HEPA vacuums were used) as compared to the lead loadings for the experiments in which the HEPA vacuum was used (precleaning geometric mean lead loading of 100 $\mu g/ft^2$). However, the post-cleaning geometric mean lead loading for the experiments in which the HEPA vacuum was used was lower than for either of the two types of experiments where non-HEPA vacuums were used. The post-cleaning geometric mean lead loading was lower for each set of experiments in which the HEPA vacuum was used. In considering these data, EPA believes that the data on the post-cleaning lead loadings are particularly important.

In assessing the performance of cleaning methods, it is not only the percent reduction that is important but also the ability to clean down to very low levels. Several studies have demonstrated that reducing lead loadings from relatively high levels to about 100 $\mu g/ft^2$ is more readily accomplished than reductions below 100 $\mu g/ft^2$ and becomes progressively harder at lower levels.

One commenter stated that EPA did not have sufficient evidence showing that HEPA vacuums are significantly better at removing lead dust than non-HEPA vacuums and cited a Canadian Mortgage and Housing Corporation study from

1992. That study was a laboratory study done in a dynamic chamber under controlled conditions and used simulated lead dust. Lead stearate, a compound not typically used in lead-based paint, was used to spike the construction dust used in the experiments. This study has various limitations.

It focused on how much of the quantity of leaded dust applied to a surface was present in the vacuum bag after vacuuming. There was no assessment of the size of the dust particles collected. Most importantly, the study did not measure the quantity of leaded dust that remained on the floor. Without this data, the efficacy of the non-HEPA vacuums cannot be assessed. In addition, the study is not very informative as to what will occur under real-world conditions.

Two years later, the same group studied 20 test rooms where they produced lead-containing dust by power sanding walls of known lead levels. Four cleaning methods were used, of which only two produced acceptable results. The two cleaning methods that did not produce acceptable cleanups were: (1) dry sweeping the floor with a corn broom followed by vacuuming with a utility vacuum; and (2) vacuuming the floor with a household vacuum cleaner followed by wet mopping with a commercial household cleaner. The other two methods that achieved cleanups resulting in floors that passed dust clearance testing were: vacuuming the floor with a utility vacuum followed by wet mopping with a 2% solution of a commercial lead-cleaning product, followed by a rinse with clean water; and vacuuming with a HEPA vacuum, followed by wet mopping with trisodium phosphate, followed by a clean-water rinse, followed by more vacuuming with a HEPA vacuum. The report concluded that these practices may not be effective.

The same commenter also referred to a report submitted to HUD by the California Department of Health Services. This study evaluated a range of vacuums. The efficacy of the non-HEPA vacuums varied, particularly in comparison with the HEPA vacuums. The authors of the report did not identify the attributes of the non-HEPA vacuums that were instrumental in determining their effectiveness. At best, vacuums that were effective at picking up and retaining lead-based paint dust could be classified as high performing although there were no criteria that could be discerned on what made a high-performing vacuum. The report also states that HEPA models without floor tool brushes performed poorly. This may be the case. The HEPA vacuums used in EPA's Dust Study performed adequately and all of these vacuums were equipped with flip-down brushes on the floor tool.

The California report contained another finding of interest. "Of special concern is the direct observation under the scanning electron microscope of lead dust particles dissolving on exposure to water to release large numbers of sub-micron lead particles. Although requiring further study, this effect suggests that vacuuming to remove most of the water soluble lead dust, followed by wet washing would be the best cleaning strategy." The cleaning protocol in this final rule follows this strategy by requiring, for all surfaces in and around the work area except for walls, HEPA vacuuming, followed by wet wiping or wet mopping, followed by the cleaning verification protocol.

EPA has determined that the weight of the evidence provided by these studies demonstrate that the HEPA vacuums consistently removed significant quantities of lead-based paint dust and reduced lead loadings to lower levels then did other vacuums.

While there may be some vacuum cleaners that are as effective as HEPA vacuums, EPA has not been able to define quantitatively the specific attributes of those vacuums. That is, EPA is not able to identify what criteria should be used to identify vacuums that are equivalent to HEPA vacuums in performance.

The authors of the studies just discussed do not state that the vacuums used are representative of all vacuums nor do they try to identify particular aspects of the non-HEPA vacuums. Thus, EPA does not believe that it can identify in this final rule what types of vacuums can be used as substitutes for HEPA vacuums. EPA believes it would be ineffective to identify specific makes or models of vacuums (e.g., the ones used in the studies) in this final rule given how quickly manufacturers change models, nor would that take into account new manufacturers.

EPA also disagrees with the commenter who suggested that vacuums retrofitted with a HEPA filter should be considered sufficient for the purposes of this rule. These vacuums are not necessarily properly sealed or designed so that the air flow goes exclusively through the HEPA filter. EPA agrees with the commenter who stated that HEPA vacuums are vacuums that have been designed for the integral use of HEPA filters, in which the contaminated air flows through the HEPA filter in accordance with the instructions of its manufacturer and for which the performance standard for the operation of the filter is defined. EPA also agrees with those commenters who contended that the rule should contain a more-specific definition of HEPA vacuum. Accordingly, this final rule defines "HEPA vacuum" as a vacuum that has been designed with a HEPA filter as the last filtration stage and includes a description of what the term HEPA means. The definition of "HEPA vacuum" also specifies that the vacuum cleaner must be designed so that all the air drawn into the machine is expelled through the filter with none of the air leaking past it.

Furthermore, EPA agrees that OSHA's requirement for HEPA vacuums should be an important consideration in determining whether HEPA vacuums should be required to be used as part of the work practices being finalized today. Because OSHA's standard covers practically all work subject to the EPA's final Renovation, Repair, and Painting Rule, and applies to all firms having an employee/employer relationship with few exceptions, there is no reason to create a separate standard for those firms not subject to the OSHA standard, particularly in light of the data on the efficacy of HEPA vacuums versus non-HEPA vacuums discussed above. Even if EPA were able to define vacuums that were acceptable substitutes to HEPA vacuums, it is not clear that the benefits would outweigh the complications associated with creating an EPA standard that is different than that required by OSHA.

Cleaning Verification

This final rule requires the certified renovator to use disposable cleaning cloths after cleaning both as a fine-cleaning step and as verification that the containment and cleaning have sufficiently cleaned up the lead-paint dust created by the renovation activity. Cleaning verification's usefulness is based on the combination of its fine-cleaning properties and the fact that it provides feedback to the certified renovator on the effectiveness of the cleaning. Cleaning verification is an important component of the work practices set forth in this rule and contributes to the effectiveness of the combination of training, containment, cleaning, and verification at minimizing exposure to lead-based paint hazards created during renovation, remodeling, and painting activities. Ongoing monitoring is needed to confirm that cleanup procedures have worked.

Disposable Cleaning Cloth Study

The Disposable Cleaning Cloth Study used commercially available disposable cleaning cloths to determine whether variations of a "white glove" test could serve as an effective alternative. White disposable wet and dry cleaning cloths were used to wipe windowsills and floors, then they were examined to determine whether dust was visible on the cloth. This determination was made by visually comparing the cloth to a photographic standard that EPA developed to correlate to a level of contamination that is at or below the dust-lead hazard standard in 40 CFR 745.65(b). Cloths that matched or were lighter than the photographic standard were considered to have achieved "white glove."

This series of studies found that on uncarpeted floors, 91.5% of the surfaces that achieved "white glove" using only dry cloths were confirmed by dust-wipe sampling to be below the dust-lead hazard standard for floors, while 97.3% of the floors that achieved "white glove" using only wet cloths were also below the hazard standard. In addition, 10 of the 11 floors where "white glove" was not achieved using dry cloths, and 20 of the 21 floors where "white glove" was not achieved using wet cloths, were nonetheless below the dust-lead hazard standard. There were very few instances where "white glove" was achieved but the dust lead level was above the dust-lead hazard standard. Thus, the study showed that for floors, the white glove test results were biased toward false positives.

Windowsills were also tested. For the dry cloth protocol, 96.4% of the sills that achieved "white glove" were also confirmed by dust wipe sampling to be below the dust-lead hazard standard for windowsills, and the one sill that did not achieve "white glove" was also below the standard. For the wet cloth protocol, all of the sills that achieved "white glove" were also below the dust-lead hazard standard, as were the four sills that did not reach "white glove."

Based on the results of the Disposable Cleaning Cloth Study, the 2006 Proposal included for interior renovations, as part of the work practices, a post-renovation cleaning verification process that would follow the visual inspection and cleaning. Cleaning verification would consist of wiping the interior windowsills and uncarpeted floors with wet disposable cleaning cloths and, if necessary, dry disposable cleaning cloths, and comparing each to a cleaning verification card developed and distributed by EPA.

The Dust Study

The Dust Study, which is described elsewhere in the preamble, assessed the proposed work practices. As one component of the proposed work practices, the cleaning verification was evaluated in the Dust Study. It should be noted that the Dust Study was not designed specifically to evaluate the cleaning verification in isolation of the rest of the work practices. Unlike the earlier Disposable Cleaning Cloth Study that was intended to test the effectiveness of the use of the "white glove" test in isolation, the Dust Study was meant to evaluate the effectiveness of the proposed work practices, including cleaning verification.

Unlike the earlier Disposable Cleaning Cloth Study, the Dust Study involved actual renovations that were performed by local renovation contractors who received instruction in how to perform cleaning verification and then were left alone to determine whether cleaning cloths matched or were lighter than the cleaning verification card. To maximize the information collected about cleaning verification during the Dust Study, cleaning verification was conducted after each experiment, not just those experiments that were being conducted in accordance with the proposed rule requirements for containment and cleaning.

One of the Dust Study conclusions was that cleaning verification resulted in decreases in lead levels, but was not always accurate in identifying the presence of levels above the Environmental Protection Agency's dust-lead hazard standards for floors and sills. This refers to the experiments involving power planing and high temperature heat guns. An examination of the cleaning verification data in the study shows that, if power planing and high-temperature heat gun experiments are excluded, the values for post-renovation cleaning verification when the proposed rule work practices were used were at or below the regulatory hazard standard for floors, often significantly below the standard. These results were similar for windowsills.

Excluding power planing and high temperature heat gun experiments, all of the post-renovation cleaning verification windowsill sample averages for experiments conducted in accordance with the proposed rule requirements were below the regulatory dust-lead hazard standard for windowsills. In addition, 26 of the 30 other experiments (using only some elements of the proposed containment and cleaning requirements) not involving power planing or high temperature heat guns had post-renovation cleaning verification sill sample averages well below the hazard standards.

Cleaning Verification as an Alternative to Clearance Testing

In determining whether cleaning verification could be seen as a qualitative alternative to clearance testing, EPA considered both the Disposable Cleaning Cloth Study and the Dust Study. Even though the Disposable Cleaning Cloth Study showed that the cleaning verification cloths that reached "white glove" were approximately 91% to 97% likely to be below the regulatory hazard standard, EPA believes the greater variability seen in the Dust Study, particularly in the experiments where the complete suite of proposed work practices were not used, does not support the characterization of cleaning verification as a direct substitute for clearance testing.

Cleaning verification, when used apart from the other work practices, is not as reliable a test for determining whether the hazard standard has been achieved as clearance testing. However, the Dust Study supports the validity of cleaning verification as an effective component of the work practices. The cleaning and feedback aspects of cleaning verification are important to its contribution to the effectiveness of the work practices.

Based on a review of the Dust Study and the Disposable Cleaning Cloth Study, EPA concluded that if the practices prohibited in this final rule are avoided and the required work practices are followed, then cleaning verification is an effective component of the work practices. EPA believes that the suite of work practices as a whole are effective at addressing the lead-paint dust that is generated during renovation, repair, and painting preparation activities.

Therefore, the final rule does not require dust clearance sampling after any renovations, nor does it allow the signs delineating the work area to be removed based solely on the results of a visual inspection. The final rule does require a certified renovator to perform a visual inspection to determine whether dust, debris, or residue is still present in the work area, and, if these conditions exist, they must be eliminated by recleaning and another visual inspection must be performed. In addition, the rule requires that after an interior work area passes the visual inspection, the cleaning of each windowsill and uncarpeted floor within the work area must be verified, as explained in the following.

After an exterior work area passes the visual inspection, the renovation has been properly completed. In response to one commenter who was concerned about the dust that could collect on exterior windowsills during exterior projects, the final rule clarifies that the visual inspection must confirm that no dust, debris, or residue remains on surfaces in and below the work area, including windowsills and the ground.

For interior renovations, after the work area has been cleaned and has passed a visual inspection, a certified renovator must wipe each interior windowsill in the work area with a wet disposable cleaning cloth and compare the cloth to a cleaning verification card developed by EPA. If the cloth matches or is lighter than the image on the card, that windowsill has passed the post-renovation cleaning verification. If the cloth is darker than the image on the card, that windowsill must be recleaned in accordance with §745.85(a)(5)(ii)(B) and (C) and the certified renovator must wipe that windowsill with a new wet cloth, or the same one folded so that an unused surface is exposed, and compare it to the cleaning verification card. If the cloth matches or is lighter than the card, that windowsill has passed. If not, the certified renovator must then wait for one hour after the surface was wiped with the second wet cleaning verification cloth or until the surface has dried, whichever is longer. Then, the certified renovator must wipe the windowsill with a dry disposable cleaning cloth. Based on the Dust Study, EPA concluded that this process need not be repeated after the first dry cloth. At that point, that windowsill has passed the post-renovation cleaning verification process. Each windowsill in the work area must pass the post-renovation cleaning verification process.

The cleaning verification protocol in the final rule is similar to what was in the 2006 Proposal. By not requiring the surface to be recleaned after the second wet wipe and by ending the cleaning verification process after one dry cloth, this final rule is different from the Proposal. The 2006 Proposal required that the dry cloths be used until one passed verification (i.e., reached "white glove"). EPA's final rule does not require more than one dry cloth because only 3 experiments out of the 60 performed in the Dust Study failed the second wet cloth. None of these 3 experiments were performed in accordance with the requirements of this final rule; all experiments performed in accordance with the requirements of this final rule passed after either the first or second wet cloth.

Based on the Dust Study, it is unlikely that dust containing lead will remain in excess of the hazard standard following two wet and one dry wipes; however EPA is concerned about the possibility of requiring potentially indefinite cleaning by renovation contractors, with the potential of making them responsible for cleaning up preexisting dirt or grime, whether lead-contaminated or not.

After the windowsills in the work area have passed the post-renovation cleaning verification, a certified renovator must proceed with the cleaning verification process for the floors and countertops in the work area. A certified renovator must wipe no more than 40 ft^2 of floor or countertop area at a time with a wet disposable cleaning cloth. For floors, the renovator must use an application device consisting of a long handle and a head to which a wet disposable cleaning cloth is attached. If the floor and countertop surfaces in the work area exceed 40 ft^2, the certified renovator must divide the surfaces into sections, each section being no more than 40 ft^2, and perform the post-renovation cleaning verification on each section separately. If the wet cloth used to wipe a particular section of surface matches or is lighter than the image on the cleaning verification card, that section has passed the post-renovation cleaning verification. If, however, on the first wiping of a section of the surface, the wet cloth does not match and is darker than the image on the cleaning verification card, the surface of that section must be recleaned in accordance with §745.85(a)(5)(ii)(B) and (C).

After recleaning, the certified renovator must wipe that section of the surface again using a new wet disposable cleaning cloth. If the second wet cloth matches or is lighter than the image on the cleaning verification card, that section of the floor has passed.

If the second wet cloth does not match and is darker than the image on the verification card, the certified renovator must wait for 1 hour or until the surface has dried, whichever is longer. Then, the certified renovator must wipe each of those 40 ft^2 sections of the floor or countertop surfaces that did not achieve post-renovation cleaning verification using the wet cloths with a dry disposable cleaning cloth. On floors, this wiping must also be performed using an application device with a long handle and a head to which the dry cloth is attached. At that point, the floors and countertops have passed the post-renovation cleaning verification process and the warning signs may be removed.

In finalizing the work practices in this final rule, EPA has taken into consideration safety, reliability, and effectiveness. EPA has concluded that these work practices, including cleaning verification, are an effective and reliable method for minimizing exposure to lead-based paint hazards created by the renovation, both during and after the renovation.

Dust Clearance Testing and Clearance

Many commenters asserted that the rule should require dust clearance testing instead of the cleaning verification. Some further contended that dust clearance testing is the only proven method for verifying lead dust levels. Others supported the use of dust wipe clearance testing for purposes of clearance for the renovation. One commenter noted that even when dust clearance testing is performed it is not uncommon for clearance to be conducted up to three times on a home to make sure that lead levels are sufficiently low. Some commenters suggested that cleaning verification be used as a screen before dust clearance testing. Other commenters contended that dust clearance testing should not be required because it is expensive and time consuming and is an obstacle to completing the renovation job. Other commenters contended that dust clearance testing has been done in some jurisdictions quickly and relatively inexpensively.

A few commenters contended that EPA should not require dust clearance testing because there is a difference between abatement, which is intended to eliminate lead-based paint hazards, and renovations in which the focus should be to not create any new lead-based paint hazards. Some commenters asserted that dust clearance testing should not be required because this would result in the renovator being responsible for existing lead-based paint hazards. One commenter used the example of a window replacement project to illustrate this point. The commenter argued that, where the floor in the work area is in poor condition but outside the scope of the renovation contract, the window replacement contractor should not be responsible for making sure the floor passes a clearance standard, which may not be possible without modifying the floor.

EPA disagrees that dust clearance testing and clearance should be components of the renovation activities subject to this final rule. Dust clearance testing is used in abatement to determine whether lead-based paint hazards have been eliminated. This test is part of a specific process that involves a specialized workforce (e.g., inspector, risk-assessor), typically removal of residents, and modifications to the housing in some instances to eliminate lead-based hazards (e.g., removing carpet or refinishing or sealing uncarpeted floors). Dust clearance testing is needed to determine if lead-based paint hazards have been eliminated and residents can reoccupy a house and not be exposed to lead-based paint hazards. As noted by a commenter, a home may require clearance testing be conducted up to three times before the home is determined to be free of lead-based paint hazards and it may require that floors be refinished or that carpets be replaced.

The Disposal Cleaning Cloth Study showed that wet wipes can pick up accumulated grime from floors. Applying this to the renovation context, if EPA were to require clearance, renovators might be held responsible for cleaning up preexisting lead dust hazards that had accumulated in the grime on the floor. Based on the Dust Study, EPA has determined that all of the leaded dust generated by the renovation will have been cleaned up by two wet wipes followed by one dry wipe, where necessary. EPA is concerned about the possibility of requiring potentially indefinite cleaning by renovation contractors, with the potential of making them responsible for cleaning up preexisting dirt or grime, whether lead-contaminated or not. Even assuming EPA has authority to require replacement of carpets and floors under some circumstances as part of a renovation project, EPA does not think as a policy matter that such an approach in which preexisting hazards must be eliminated is appropriate. It could fundamentally change the scope of a renovation job.

The time and cost of conducting clearance testing and achieving clearance is an acceptable part of the time and cost of conducting the abatement given the goal of an abatement, the range of activities that are inherent in an abatement, and the activities that are required to be conducted to achieve clearance. Given the effectiveness of the work practices being finalized in this rulemaking, including the role of cleaning verification in minimizing exposure to lead-based paint dust generated during renovations, dust clearance testing does not provide the added value to balance the time and effort and the cost to home and building owners associated with requiring this additional step to the work practices.

As discussed in Unit II.A.6.b., there are many differences between renovations and abatements. Renovations are different from abatements in intent, implementation, type of workforce, workforce makeup, funding, and goal. Renovations are focused not on eliminating lead-based paint hazards, but rather on making repairs or improvements to a building. The vast majority of abatements are done with funding from HUD and/or a state or local government. In addition, residents are not typically present in a residence during an abatement while they are typically present in a residence during a renovation. Thus, the purpose of dust wipe clearance testing and clearance would necessarily be different if it were used in a renovation rather than in an abatement.

For abatements, clearance testing and clearance are used to minimize potential exposure by eliminating lead-based paint hazards after completion of the job. Clearance acts as the means to ensure that minimization, and signal the end of the job. For renovations, given the presence of residents, the concern is for potential exposure both during and after the job. Dust clearance testing and clearance would only address the second part of the exposure equation. Thus, dust clearance testing conducted after renovation activities have been completed would not provide the equivalent determination of potential exposure that it does for abatement.

EPA has considered this difference as one factor in its determination that given the effectiveness of the work practices being finalized in this rulemaking, including the role of cleaning verification in minimizing exposure to lead-based paint dust generated during renovations, dust clearance testing does not provide the added value to balance the time and effort and the cost to home and building owners associated with requiring this additional step to the work practices.

Although renovators should be required to address lead-based paint dust generated by renovation activities, the agency is not requiring renovators to take the actions required under the abatement rules to achieve clearance for lead-based paint dust not associated with the renovation and to address housing conditions not associated with the renovation.

EPA agrees that having dust wipe samples collected by a qualified person and analyzed by a qualified laboratory is an effective way to determine the quantity of lead in dust remaining after a renovation activity, but it would not necessarily show that the dust was due to the specific renovation activity. EPA also notes that in addition to providing a numerical value, dust clearance testing costs more than cleaning verification and takes longer to produce results. Results can take from 24 to 48 hours or longer and cleaning, sampling, and analysis may have to be repeated depending on the initial results. During this period, the warning signs delineating the work area would need to be maintained to protect occupants and others from the risk of exposure to lead-based paint hazards created by the renovation.

Thus, EPA believes that dust clearance sampling is a poor fit for renovation work for a variety of reasons, including the greater expense associated with clearance testing, the time necessary to obtain the results of the testing and the consequent delay in the completion of the job, and the potential to expand the scope of the renovation.

EPA believes that dust clearance testing and clearance are not necessary given that the Dust Study demonstrates that cleaning verification, as an effective component of the work practices, minimizes exposure to lead-based paint hazards created by the renovation, both during and after the renovation. The cleaning and feedback aspects of cleaning verification are important to its contribution to the effectiveness of the work practices. EPA notes that unlike dust wipe clearance testing in which a small part of the work area would be tested, cleaning verification is conducted over the whole work area. Each repetition of the cleaning verification protocol further cleans the surface.

The work practices, including cleaning verification, required by this final rule are expected to minimize exposure to any newly created lead-based paint hazards created by a renovation by removing newly deposited dust, while requiring cleanup of preexisting hazards only incidentally, to the extent such cleanup is unavoidable to address the newly created hazards. The Dust Study demonstrates that the cleaning verification protocol, used in conjunction with the other work practices in this final rule, is effective and reliable in achieving this result.

While the requirements of this rule will, in some cases, have the ancillary benefit of removing some preexisting dust-lead hazards, it strikes the proper balance of addressing the lead-based paint hazards created during the renovation but at the same time not requiring renovators to remediate or eliminate hazards that are beyond the scope of the work they were hired to do.

▶ VISUAL INSPECTION IN LIEU OF CLEANING VERIFICATION

Some commenters urged EPA to require only visual inspection of the work area after the cleaning following a renovation. They contend that cleaning verification is not needed. Some commenters argued that thorough cleaning in combination with a requirement that no visible dust or debris remain is adequate to address the lead dust created by the renovation activity. Most of these commenters also noted that because renovation and abatement are different, it would be inappropriate for EPA to impose additional requirements on renovation firms beyond visual inspection. Some commenters contended that the lead dust from a renovation is usually in the form of debris such as chips and splinters that can be seen with the naked eye, and the presence of this debris is an indicator to workers that the job site requires additional cleaning until no visible debris remains.

One commenter contended that cleaning after the renovation activity until the worksite passed a visual inspection was the most important determinant of whether a job would pass a dust clearance test. In support of this contention, the commenter cited the Reissman study (*Journal of Urban Health*, 79(4):502-11, 2005). The commenter contended that the study demonstrates that when there was no visible dust and debris present after completion of renovation or remodeling activity, there was no added risk of a child having an elevated blood-lead level as compared to the risk for children living in homes where there was no reported renovation or remodeling work.

Two commenters offered an analysis of two sets of data collected by an environmental testing firm. One dataset consists of post-renovation dust samples collected in Maryland apartment units; the other consists of dust samples collected for risk assessment purposes in 41 states. No information on renovation activity is provided for the second dataset. The commenters argue that because 96.7% of the Maryland post-renovation samples and 96.1% of the other samples were below the applicable hazard standard for the surface (floor or windowsill) tested, this suggests that visual inspection in those cases was sufficient to ensure that no dust-lead hazard existed.

One commenter cited the Dust Study, the NAHB Lead-Safe Work Practices Survey, and several other studies as supporting the conclusion that lead-safe work practices and modified lead-safe work practices, along with a two-step or three-step cleaning process using a HEPA-equipped vacuum and wet washing, greatly reduce dust lead levels and should be regarded as best management practices for renovation jobs. The commenter notes that the NAHB study found significant reductions in loading levels after cleanup using HEPA-equipped vacuum and then either wet washing or using a wet-mopping system. The commenter argues that if the work area is cleaned using these practices, it is appropriate to adopt a visual clearance standard allowing no visible dust or debris in the work area at the conclusion of the job.

Other commenters contended that visual inspection following cleaning after a renovation is not a reliable method for determining whether a lead-based paint hazard remains after cleaning. Some commenters cited a study conducted by the National Center for Healthy Housing (NCHH) showing that 67% of the visual inspections that initially passed failed when checked more carefully and 54% that eventually passed a visual inspection were found to be above the hazard standard. However, one commenter contended this was a poorly conducted study. Another commenter referred to the study "An Evaluation of the Efficacy of the Lead Hazard Reduction Treatments Prescribed in Maryland Environmental Article 6–8" conducted by NCHH for the Baltimore City Health Department in which 53% of housing identified by visual inspection as being below the hazard standard was actually above the hazard standard. Another commenter argued that NIOSH research indicates that significant lead contamination may remain on surfaces that appear clean.

During interagency review, one commenter pointed to 2007 studies from Maryland and Rochester, New York that they contend show trained workers and visual inspection for dust and debris can achieve 85 to 90% compliance with the hazard standards following renovations in previously occupied housing. Given the lateness of the submission, EPA did not review this information. However, EPA notes that in a cover letter, the commenter states that the 2007 Maryland study was conducted by workers who had taken a 2-day training course, which is more training than required by this rule. Even if the studies do demonstrate this effectiveness by highly trained workers, EPA does not believe that an 85 to 90% effectiveness is sufficiently protective for residents.

EPA disagrees with those commenters who contended that a visual inspection following cleaning after a renovation is sufficient to ensure the lead-based paint dust generated by a renovation has been sufficiently cleaned-up. The weight of the evidence clearly demonstrates that visual inspection following cleaning after a renovation is insufficient at detecting dust-lead hazards, even at levels significantly above the regulatory hazard standards. Further, EPA disagrees with the implication that easily visible paint chips and splinters are necessarily the primary materials generated during a renovation. EPA studies, including the Dust Study, show that renovation activities generate dust as well as chips and splinters.

Finally, EPA disagrees with those commenters who requested the work practices in this final rule not include any verification beyond visual inspection. In the Dust Study, there were 10 renovations performed in accordance with the 2006 proposed work practices that did not involve practices prohibited by this final rule. Of those 10 renovations, 5 needed the additional cleaning verification step to achieve EPA's regulatory dust-lead hazard standards for floors. (EPA notes that the Dust Study Protocol did not explicitly specify that all dust and debris be eliminated prior to the cleaning verification step, only that visible debris be removed. However, the contractor running the study for EPA reported that, in practice, the renovators participating in the study eliminated all visible dust and debris as part of their typical cleaning regimen. Thus, the study protocol was slightly different from the rule requirements, which state that the renovation firm must remove all dust and debris and conduct a visual inspection before beginning the cleaning verification procedure.)

EPA does not believe that the Reissman et al. study is supportive of the contention that visual inspection of the work area is sufficient because it did not evaluate the effectiveness of a visual inspection requirement. The study did not measure dust lead levels, which are the basis for this rule. Instead, it characterized the relationships between elevated blood-lead levels and renovation dust and debris that spread throughout the housing. EPA notes that Reissman et al. concluded that there was a correlation between renovation activities and elevated blood-lead levels.

EPA concluded that the dataset referenced by one commenter that consists of dust samples collected for risk assessment purposes in 41 states is not informative because there was no information on renovation activity collected with these dust samples. With respect to the Maryland renovation study, 96.7% is an overstatement. The author who conducted the analysis stated that when the maximum test values are examined rather than the mean, 9.8% of the Maryland sample and 12.5% of the national sample of properties with LBP surpassed at least one of the hazard thresholds of 40 µg/sf for floors and 250 µg/sf for sills. One of the study's exhibits showed a fairly sizable percentage of the lead tests exceeded the clearance thresholds. The failure rates were about 20% lower for Maryland than for the national LBP sample. However, even for Maryland, nearly one in ten apartments would fail the hazard test. Thus, even if these were the only data available, it would not support the conclusion that visual clearance is effective.

After reviewing the NAHB Lead-Safe Work Practices Survey, EPA concluded that it does not support the contention that visual inspection is sufficient to detect whether lead-based paint dust remains. While EPA agrees that use of a HEPA vacuum and wet washing are effective at cleaning lead-based paint dust, this does not support the case for relying on visual inspection without subsequent cleaning verification. In the NAHB study, the levels of lead-based paint dust that remained after the renovation activities were sometimes higher and sometimes lower than at the start of the renovation, but they were always at relatively high levels after the renovation—as high as 11,400 $\mu g/ft^2$. In addition, the two studies conducted by the National Center for Healthy Housing as noted by commenters demonstrate that visual inspection was not effective at determining the presence of dust-lead hazards. The Evaluation of the HUD Lead-Based Paint Hazard Control Grant Program study conducted by NCHH corroborates these findings.

Some commenters were concerned that cleaning verification is not intended for use on carpeted floors. They were not confident that thorough cleaning was adequate to address potential lead hazards that might remain in carpet after the renovation. One commenter pointed to studies showing a significant correlation between dust lead in carpets and children's blood lead. As cleaning verification is not required for carpet, commenters criticized the lack of a required method for determining that lead hazards in carpet had been eliminated. Commenters suggested EPA require clearance testing for carpeted rooms in the work area, which some argued has been demonstrated to be effective, or rely on the HUD protocol, which they asserted is widely accepted and used.

As discussed in detail in Unit IV.E. of the preamble to the 2006 Proposal, EPA did not design cleaning verification for use on carpeted floors. This was based on EPA's concerns about the validity of dust wipe sampling on carpeted floors. EPA noted that the decision to apply the clearance standard promulgated in the TSCA section 403 rulemaking to carpeted floors ultimately had little consequence, given the context in which clearance standards are used—to ensure that lead-based paint hazards have been eliminated. Typically, during an abatement, carpets that are in poor condition or are known to be highly contaminated are removed and disposed. EPA further notes that the HUD Lead-Safe Housing Rule only requires HEPA vacuuming, not steam cleaning or shampooing.

While an abatement might require the removal of a lead-contaminated carpet, EPA has concluded that it is not appropriate to require carpet removal following a renovation. Even assuming EPA has authority to require removal of carpet following a renovation, this could significantly expand the cost of a renovation, and fundamentally expand the scope of the renovation activity contracted by the homeowner or building owner by requiring removal of carpets as a result of preexisting lead contamination.

Dust Study data on containment and information on the effectiveness of HEPA vacuums show that the use of containment and post-renovation cleaning with HEPA vacuums to remove the lead-based paint dust potentially deposited on the carpets during the renovation would reliably and effectively address lead-based paint dust generated during a renovation. Thus, rather than rely on a dust clearance sample that may not be accurate and may require the replacement of the carpet for renovation projects in which a carpet is present, EPA is finalizing the work practices that require containment and the use of a HEPA vacuum equipped with a beater bar for cleaning.

In the absence of a practical, effective way of determining how much lead dust has been added to a carpet and whether it has been fully removed, EPA is adopting a technology-based approach for carpets that differs from the approach used for hard-surfaced floors, by requiring use of a HEPA vacuum with a beater bar. EPA is not aware of, and commenters have not identified, a practicable approach similar to the one EPA has adopted for floors as a basis to evaluate the results of the application of work practice standards to carpets. In the absence of such an approach, EPA believes the approach adopted in the final rule is the most effective, reliable approach available for minimizing potential lead-based paint hazards in carpets created by renovations.

One commenter suggested that cleaning verification be required on other horizontal surfaces within the work area, in addition to windowsills and uncarpeted floors. EPA agrees with this commenter because the Dust Study demonstrated that, in nearly all cases, the cleaning verification step resulted in lower dust lead levels and, in most cases, the verification step was needed to achieve cleanup of all of the leaded dust deposited on the floors by the renovation. EPA is also concerned about the possible contamination of surfaces that are used to prepare, serve, and consume meals. EPA expects that movable surfaces, such as tables and desks, will be moved from the work area before work begins. Therefore, EPA has modified the rule to require cleaning verification on all countertops in the work area.

The Environmental Protection Agency received comments prior to the 2007 request for comments on the proposed work practices in light of the Dust Study. Those pre-Dust Study comments are summarized in the following paragraphs.

Commenters questioned whether cleaning verification had been demonstrated to be valid, reliable, effective, or efficient in establishing that the work area had been adequately cleaned or that the clearance standards were met. Some commenters contended that the cleaning verification method showed promise, but should be subjected to additional testing, including field trials, to demonstrate its effectiveness when used by certified renovators. Commenters on the 2006 Proposal observed that the cleaning verification protocol was supported by a single study that was conducted under conditions unlike those presented by the typical renovation. Specifically, a commenter noted that most of the housing units studied had undergone some form of abatement that would likely have reduced dust levels and the study used professional inspectors or other highly trained individuals to collect the samples according to specified protocols. The commenter was concerned that a renovator with no experience with sample collection and little training could replicate the work of the professionals used in the study. The commenter pointed out that the study avoided testing the procedure on rough surfaces, a condition that will frequently occur in real-world applications, and used a different set of wipe protocols than actually utilized by EPA in the 2006 Proposal. Another commenter on the 2006 Proposal noted that cleaning verification had never been employed in a real-world practical setting. In addition, some of these commenters contended that the cleaning verification protocol was too complicated or too confusing to follow.

A number of commenters who provided comments in response to EPA's request for comments on the proposed work practices in light of the Dust Study quoted the sentence in the conclusion section of EPA's Dust Study that states that the cleaning verification protocol was not always accurate in identifying the presence of levels above EPA standards for floors and sills. Some of these commenters also noted the Dust Study report's discussion of factors that affected the effectiveness of cleaning verification, such as floor condition, contractor performance, job type, and dust particle characteristics. One commenter observed that while all interior experiments resulted in final passed cleaning cloths for all floor zones and for all windowsills, nearly half of the experiments in the study ended with average work room floor lead levels above EPA's dust-lead hazard standard for floors of 40 $\mu g/ft^2$. The Clean Air Scientific Advisory Committee, while not asked to comment on the efficacy of the cleaning verification, contended that in the Dust Study cleaning verification did not provide sufficiently reliable results, leading to an inaccurate assessment of cleaning efficiency.

EPA disagrees with these commenters. The Dust Study did provide a real-world practical setting in which to assess the use of cleaning verification. Local renovation contractors performed actual renovations for each experiment in the study. The contractors performed cleaning verification on floors of wood, vinyl, or tile, in good, fair, or poor condition. The Dust Study used the protocols that were consistent with those in the 2006 Proposal.

While the Dust Study was not designed specifically to assess cleaning verification, it did assess the effectiveness of cleaning verification both when it was used as part of the proposed rule work practices and as a separate step after the other experiments that did not follow all the proposed work practices. Each experiment included a cleaning verification step. The contractors were instructed in how to perform cleaning verification. They independently determined whether particular cloths matched or were lighter than the cleaning verification card. In most renovations not involving the practices that EPA is prohibiting in this rule, that is, power planing, power sanding, and the use of high temperature heat guns, cleaning verification in combination with the other work practices were effective at reducing dust lead levels on surfaces at or below the dust-lead hazard standards, regardless of the condition of the floor. Cleaning verification, as well as the other components of the work practices that were finalized were not effective when high dust generation practices, such as power planing (including power sanding) and high temperature heat guns, were used. These practices, as well as torching, are prohibited in this rulemaking. Thus, EPA, in its determination on the effectiveness of cleaning verification, is focusing on the results of the experiments in the Dust Study that did not involve these prohibited practices.

Of the 10 experiments in which the proposed rule practices were used and in which the practices prohibited in this final rule were not used, all final lead-based paint dust levels were at or below the regulatory hazard standard (taking into account the accepted level of uncertainty—that is, within plus or minus 20%, which is the performance criteria for the National Lead Laboratory Accreditation Program. In fact, four experiments resulted in levels that were less than 10 $\mu g/ft^2$, three resulted in levels less than 30 $\mu g/ft^2$, and three resulted in levels that were approximately 40 $\mu g/ft^2$ (all were well within the level of uncertainty for this value). In four of the experiments, at least one floor area failed verification on the first wet disposable cleaning cloth, all passed on the second wet cloth.

In one of the experiments, a windowsill failed the first wet cloth, but passed the second. These results were seen on floors in a variety of conditions, including good, fair, and poor conditions. As a general case, in the other experiments that did not follow all the proposed work practices, the use of cleaning verification after cleaning (both baseline cleaning and cleaning following the proposed work practices) reduced, often significantly, the amount of lead dust remaining.

EPA agrees with commenters that cleaning verification should not be used for clearance. However, while cleaning verification is not clearance testing, as described above, the use of cleaning verification consistently resulted in levels of lead-based paint dust at or below the hazard standard. Also, the use of cleaning verification consistently resulted in lower levels of lead-based paint dust than remained after all types of cleaning studied when only followed by visual inspection. There is sufficient consistency in the data to support the use of cleaning verification as an effective component of the work practices that were finalized.

In response to the comment that the Disposable Cleaning Cloth Study used professional inspectors or other highly trained individuals following specified protocols, EPA intends to include cleaning verification in its training course for renovators and will use the results of the Dust Study and the agency's observations on the experience of the contractors in the study in its development of this course.

Many commenters objected to the "white glove" standard as inherently subjective, and doubted whether it would be protective. The commenters were concerned that the effectiveness of cleaning verification relies on a renovation worker's understanding and application of the protocol, ability to define the floor sampling area or areas, and use of the cleaning verification card to determine whether a surface has been adequately cleaned. One commenter contended that, based on its experience as a subcontractor to EPA on the Disposable Cleaning Cloth Study, making the visual pass/fail determination can be quite subjective and open to interpretation. The commenter believes that it may be unrealistic to expect that renovation workers will consistently make the proper decision using the proposed verification card.

Some commenters speculated that the renovator's accuracy in comparing the cleaning cloth to the verification card could depend on factors such as the renovator's visual acuity, the lighting in the room, or simply differences in judgment among renovators. Another commenter thought that the lack of corrections for surface conditions, the experience of the person conducting the visual assessment, or preexisting conditions might bias the results of testing.

EPA agrees that visual comparison of a cleaning cloth to a cleaning verification card has an element of subjectivity because the visual comparison of cloth to card requires some exercise of judgment on the part of the person doing the comparing. However, this does not necessarily mean that the comparison is suspect. As previously stated, the Dust Study represents a real-world test of the ability of renovators to learn how to do cleaning verification and to apply it in the field. Although one participant in the Dust Study expressed concern about the subjectivity of the test, the fact remains that cleaning verification was successfully performed by the renovation contractors in all of the experiments involving the work practices being finalized in this final rule (excluding those involving power planing, power sanding, and high temperature heat guns) and was predictive of whether renovators had cleaned up the lead-based paint hazards created during the renovation activity to the dust-lead standard, particularly when the proposed work practices were used. These cleaning verifications were conducted by various persons in various light conditions and on various surface conditions.

Further, EPA notes that cleaning verification is not simply qualitative clearance. Unlike the sampling for dust clearance testing, the cleaning verification involves a cleaning component. The act of doing the cleaning verification has been shown to lower, often significantly, the dust lead levels. Finally, in the development of its training course for contractors, EPA plans to use its data on the contractors' use of cleaning verification in the Dust Study, including their use of the cleaning verification cards.

Some commenters were concerned that the cleaning verification protocols are too impractical, burdensome, or time-consuming for many contractors to perform. However, the Dust Study found that cleaning verification only took, on average, slightly less than 13 minutes for experiments where the proposed rule requirements were followed. EPA's Final Economic

Analysis estimates that the average cost of cleaning verification ranges from less than $10 to $30 in residences, and in public and commercial buildings with COFs, it ranges from less than $10 to less than $50.

One commenter asked about the availability of the cleaning verification card, specifically, who would produce them, where would they be available, and how often do they need to be replaced. EPA intends to produce the cleaning verification cards and to make them available at accredited renovator training courses and on request from the National Lead Information Center. Several commenters argued that a third party should perform cleaning verification (or visual inspection, in the case of exterior jobs) rather than the certified renovator.

Commenters saw a conflict of interest, since by performing the cleaning verification the certified renovator is evaluating the effectiveness of his or her own work. Some thought the subjective nature of the method left it open to misinterpretation or fraud. Commenters were concerned that given the competitive pressures of the renovation industry and lack of independent oversight, it was not realistic to expect all renovators to follow the cleaning verification protocol in good faith. Others worried that a renovator might feel pressured to produce a passing result, perhaps to the point of recording false results. One commenter stated that those who would not comply with the cleaning procedure are unlikely to comply with cleaning verification.

Again, as described above, EPA addressed potential conflicts-of-interest in its lead-based paint program in the preamble to the final Lead-Based Paint Activities Regulations. That discussion outlined two reasons for not requiring that inspections or risk assessments, abatements, and post-abatement clearance testing all be performed by different entities. The first was the cost savings and convenience of being able to hire just one firm to perform all necessary lead-based paint activities. The second was the potential regional scarcity of firms to perform the work. EPA believes that these considerations may be equally applicable to renovations, and perhaps more compelling, given the objective of keeping this rule simple and relatively inexpensive.

EPA is concerned that a requirement that contractors engage a third party for every renovation job will add undue complication and expense to home renovations, and that it could delay completion of renovation jobs. There are estimated to be 8.4 million renovation events annually. Moreover, as stated above, it is not uncommon for regulated entities to make determinations relating to their regulated status. Thus, after weighing these competing considerations, EPA has decided to take an approach that is consistent with the approach taken in the 402(a) Lead-Based Paint Activities Regulation and not require third party visual inspections, testing, or cleaning verification.

Some commenters contend that cleaning verification is not protective because it was designed to pass based on the regulatory hazard standard for floors. These commenters contend that this level is too high to be protective and that continuing to use this level is unwarranted given more recent data that demonstrates that lead causes neurocognitive effects at levels much lower than 10 µg/dL, the current CDC blood-lead level of concern that was used in establishing the regulatory hazard standards.

EPA interprets the statutory directive to take into account safety when promulgating work practice standards as meaning that such work practice standards should be established in relation to lead-based paint hazards—as identified pursuant to TSCA section 403. There is no level of lead exposure that can yet be clearly identified, with confidence, as clearly not being associated with potentially increased risk of deleterious health effects. EPA does not believe the intent of Congress was to require elimination of all possible risk arising from a renovation, nor is EPA aware of a method that could reliably and effectively accomplish this.

Given that the hazard standards are the trigger for regulation under section 402(c)(3) and that they are set through rulemaking, EPA has concluded that it makes most sense to use the same standards as the target level for safe work practices. Otherwise, the potential is created for a scheme under which any renovation activities found not to create hazards are not regulated at all, whereas renovation activities found to create hazards trigger requirements designed to leave the renovation site cleaner than the unregulated renovations. Given the Congressional intent that the section 403 hazard standards apply for purposes of subchapter IV of TSCA, EPA is applying them as the target level for safe work practices, which include the cleaning verification process, in this rule.

Several commenters recommended that Environmental Protection Agency adopt HUD's clearance requirement for activities other than abatement, which some commenters noted has been successfully implemented in projects in federally assisted housing. One pointed out that renovators have accepted HUD's clearance testing protocol, and implementing the "white glove" method will cause confusion in the industry and give contractors a reason for not following lead-safe work practices. A commenter recommended that EPA adopt HUD's standard for exterior clearance of visual inspection of the work area and a soil test. Commenters expressed concern that the final rule could undermine more stringent state and local standards, and asked EPA to make clear that more stringent state and local requirements for clearance would apply despite the lack of mandatory clearance in the final rule.

This final regulation does not supersede more stringent or different requirements for interim control projects or renovations regulated by HUD, the states, or local jurisdictions. Renovation firms are still responsible for complying with all applicable federal, state, or local laws when conducting renovations. In some cases, this may mean that dust clearance testing must be performed at the conclusion of a renovation rather than cleaning verification. EPA believes that renovation firms will be able to integrate these new requirements into their existing business practices with very little difficulty.

EPA also notes that the scope of the housing covered by HUD is different than the scope covered by this final rule. As noted by the commenter, HUD covers activities in projects in federally assisted housing. The occupancy patterns, including turnover, will be different than in the general population covered by this final rule. While there is some overlap, there are substantial differences. Thus, EPA believes that total consistency with HUD is not needed.

EPA proposed to allow optional dust clearance sampling at the completion of renovation activities instead of the post-renovation cleaning verification described in §745.85(b). Some commenters agreed that the decision whether to perform clearance at the conclusion of the job should be left to the homeowner. One commenter asked EPA to require that, if a resident arranged for clearance testing and found lead hazards, the contractor would have to reclean to the resident's satisfaction.

As discussed, dust clearance sampling and cleaning verification are not surrogates and EPA is not requiring renovation firms to perform an abatement (i.e., eliminate all lead-based paint hazards) as part of a renovation. The Dust Study demonstrated that cleaning verification is quite often needed to minimize exposure to dust-lead hazards created during renovations. EPA is concerned that if dust clearance sampling were allowed instead of cleaning verification, without an accompanying requirement that the renovation firm reclean until clearance is achieved, the rule would actually be less protective because the surfaces in the work area could be left less clean than if cleaning verification were performed.

In response to these comments, EPA has further considered the issue and decided to allow dust clearance sampling instead of cleaning verification only in certain limited situations. EPA agrees with the commenters that, if the rule were to allow clearance sampling instead of verification, EPA would have to require the renovator to achieve clearance, otherwise there would be no check on whether the renovation had been safely performed. HUD's Lead-Safe Housing Rule requires clearance to be achieved in many situations, as do several states. For example, the State of New Jersey requires dust clearance sampling and clearance in certain situations in multi-unit rental housing.

As noted in Unit III.G of this preamble, states, territories, and tribes may choose to have as protective as or more protective requirements than this final rule. One example of a more protective requirement would be a requirement to perform dust clearance testing and achieve clearance after renovations. Another example may be requiring that trained renovation workers demonstrate achievement of clearance levels by other cleaning verification methods, such as using newer technologies. If a firm can demonstrate, for example, using data obtained in the field, that it regularly meets the clearance standards without using the EPA specified approach but rather by using newer technology or alternative methods, a state may request that EPA evaluate such a provision as being as protective as or more protective than the methods described in this final rule.

Therefore, in situations where the contract between the renovation firm and the property owner or another regulation, such as HUD's Lead-Safe Housing Rule or a state regulation, requires dust clearance sampling by a properly qualified person and requires the certified renovator or a worker under the direction of the certified renovator to reclean until clearance is achieved,

EPA will allow the renovation firm to use both dust clearance testing and clearance instead of the cleaning verification step.

Property owners in other situations may still choose to perform dust testing at any time, such as after a renovation has been completed, including cleaning verification. EPA recommends that property owners who choose to have dust testing performed use certified dust sampling professionals such as inspectors, risk assessors, or dust sampling technicians. EPA also recommends that property owners who wish to have dust testing performed after a renovation reach an agreement with the renovation firm up-front as to what will happen based on the results of the dust testing, such as whether additional cleaning will be performed if the surfaces do not achieve the clearance standards in 40 CFR 745.227(e)(8)(viii).

Having covered the major work practices, in the next chapter we will look at recordkeeping requirements, another base that the Lead Rule covers.

Recordkeeping Requirements

9

Recordkeeping is an essential part of any business. This is especially true when you are working with hazardous materials. There are regulations that require a certain amount of recordkeeping. Beyond those there is the need to protect yourself with paperwork. Lawsuits should be avoided whenever possible, and strong records are very valuable during a dispute. This chapter looks at the requirements for recordkeeping. At the end of this chapter are several examples of forms that you may find useful in the process. I decided to put all of the forms together to make them very accessible. First, let's learn about the rules.

EPA's section 40 CFR 745.86 already requires that persons performing renovations in target housing document compliance with the lead hazard information distribution provisions of the Pre-Renovation Education Rule. Consistent with the 2006 Proposal, this final rule replaces the existing 40 CFR 745.88 as it contains only sample acknowledgment statements for the purpose of documenting compliance with the information distribution requirements and is thus unnecessary. EPA received no comments on this proposed deletion. In addition, EPA received no substantive comments on the sample acknowledgment form provided with the proposed rule. New sample acknowledgment forms incorporating language consistent with this final rule and reflecting commenter editorial suggestions are available on EPA's website at *www.epa.gov/lead* and from the National Lead Information Center at 800-424-LEAD (5323).

In addition, as proposed in the 2006 Proposal, EPA has modified paragraph (a) of 40 CFR 745.86 to make compliance with the recordkeeping requirements the responsibility of the renovation firm, not the certified renovator. Although, as discussed in the following, this final rule requires the certified renovator assigned to a renovation to certify compliance with the work practice requirements for that renovation, the renovation firm may choose to delegate other tasks associated with recordkeeping requirements to someone other than a certified renovator. For example, this rule does not require a certified renovator to distribute lead hazard information to owners and occupants before a renovation, nor does it require a certified renovator to obtain the necessary acknowledgment statements or certified mail receipts. The renovation firm may decide that it is more efficient to have someone other than the certified renovator perform these tasks.

This final rule expands the information distribution requirements to renovations in child-occupied facilities (COFs). In proposing this expansion, the 2007 Supplemental Proposal included associated recordkeeping requirements for firms performing renovations in COFs. Although EPA did receive comments on extending the information distribution requirements to child-occupied facilities, none of these comments specifically addressed the recordkeeping provisions themselves. EPA has determined that the recordkeeping requirements are an important part of monitoring compliance with and ensuring the effectiveness of the information distribution provisions of this rule.

Therefore, this final rule retains the existing recordkeeping requirements for pre-renovation lead hazard information distribution in target housing and extends those recordkeeping requirements to renovations in child-occupied facilities. Firms performing renovations in target housing or COFs must obtain and retain signed and dated acknowledgments of receipt of the lead hazard information from building owners or a certificate of mailing for such information.

In addition, renovation firms must obtain and retain signed and dated acknowledgments of receipt from the occupant (the resident of the housing unit being renovated or the proprietor of the child-occupied facility) or certificates of mailing for such information, or the firm must prepare a certification that documents the attempts made to provide this information to the occupants. For renovations in common areas in target housing, the firm must also document the steps taken to provide information to the tenants with access to the common area being renovated.

Finally, firms performing renovations in COFs must take steps to provide information to the parents and guardians of children under age 6 using the facility. Firms may do this by either mailing each parent or guardian the lead hazard information pamphlet and a general description of the renovation or by posting informational signs where parents and guardians are likely to see them. Informational signs must be accompanied by a posted copy of the pamphlet or information on how to obtain the pamphlet at no charge to interested parents or guardians. The firm's activities with respect to parents and guardians must also be documented.

▶ DOCUMENTATION OF COMPLIANCE WITH OTHER REGULATORY PROVISIONS

This final rule provides for a number of exceptions. Unit III.A.3 of this preamble describes an exception for renovations in owner-occupied target housing that is neither the residence of a child under age 6 or a pregnant woman, nor a child-occupied facility. For a renovation to be eligible for this exception, the renovation firm must obtain a signed statement from the owner of the housing to the effect that he or she is the owner of the housing to be renovated, that he or she resides in the housing to be renovated, that no child under 6 or no pregnant woman resides there, that the housing is not a COF, and that the owner acknowledges that the work practices to be used during the renovation will not necessarily include all of the work practices contained in EPA's Renovation, Repair, and Painting Rule.

Consistent with the 2006 Proposal and the 2007 Supplemental Proposal, this final rule requires renovation firms to maintain this signed statement, which must include the address of the housing being renovated, for 3 years after the completion of the renovation. Again, although EPA received comments on the merits of this exception, no comments were directed specifically to the recordkeeping requirement. EPA has determined that the recordkeeping requirement is necessary to allow EPA to monitor compliance with the terms of this exception.

This final rule also requires firms performing renovations to retain documentation of compliance with the work practices and other requirements of the rule. Specifically, the firm must document that a certified renovator was assigned to the project, that the certified renovator provided on-the-job training for workers used on the project, that the certified renovator performed or directed workers who performed the tasks required by this final rule, and that the certified renovator performed the post-renovation cleaning verification. This documentation must include a copy of the certified renovator's training certificate. Finally, the documentation must include a certification by the certified renovator that the work practices were followed with narration as applicable. The certification must include the specific information listed in Section 745.86(b)(7). The firm must keep this information for 3 years after the completion of the renovation.

The 2006 Proposal also included a requirement that renovation firms maintain documentation of compliance with the renovator and worker training requirements and the work practice requirements. This documentation would have had to include signed and dated descriptions of how activities performed by the certified renovator were conducted in compliance with the proposed requirements. To demonstrate how these recordkeeping requirements might be met, EPA prepared and placed into the docket a draft recordkeeping checklist.

EPA received many comments on the substance of these recordkeeping requirements and on the draft recordkeeping checklist. Some commenters thought that the purpose of the recordkeeping requirement should be to provide important information to consumers or to serve as part of the record of whether a particular structure was lead-safe. Some, but not all, of these commenters suggested that there was no need for the renovation firm to retain the records it prepares. Rather, the records should be given to the owners and occupants of the building either before or after the renovation. However, as proposed, the recordkeeping requirement serves two purposes.

The first is to allow EPA or an authorized state to review a renovation firm's compliance with the substantive requirements of the regulation through reviewing the records maintained for all of the renovation jobs the firm has done. The second is to remind a renovation firm what it must do to comply. EPA envisioned that renovation firms would use the recordkeeping requirements and checklist as an aid to make sure that they have done everything they are required to do for a particular renovation. For these two purposes, there is no substitute for recordkeeping by renovation firms.

However, EPA agrees with those commenters who felt that the recordkeeping requirements were vague, particularly in light of the draft recordkeeping checklist itself and the amount of time that EPA estimated it would take a renovation firm to complete the checklist. Many commenters said that it was unclear how much detail EPA would be looking for in descriptions of how the firm complied with the various work practices, and some noted that an extensive narrative would contribute no more to compliance or enforcement than a box checked to indicate that the requirements had been complied with.

In response to these commenters, EPA has revised that draft recordkeeping checklist to be more in the nature of a checklist, with a certification that the representations on the form are true and correct. Narrative information is still required where necessary, such as an identification of the brand of test kits used, the locations where they were used, and the results. EPA has also revised the regulatory text to describe the specific information that must be provided and the specific items for which a certification of compliance is required.

The regulatory text at 40 CFR 45.86(b)(7) now contains a list of work practice elements that must be certified as having been performed. In response to two commenters who suggested that the only person truly capable of certifying that the lead-safe work practices were followed on a particular job would be the certified renovator assigned to that job, EPA is requiring the certification to be completed by the certified renovator assigned to the renovation. EPA has determined that a review of the records maintained by renovation firms will be an effective method of determining whether a particular firm is generally complying with the regulations or not.

▶ NOTIFICATION TO EPA

In the 2006 Proposal, EPA requested comment on, but did not propose, a requirement that renovation firms notify EPA before beginning a covered renovation project. Most commenters supported a notification requirement, arguing notifications would provide information to EPA about where renovation activities will be occurring, so EPA could inspect ongoing renovation projects for compliance with the requirements of this rule. These commenters stated that EPA would be unable to enforce the requirements of the rule without a notification provision. Some commenters also suggested that the act of informing EPA of their activities provides a powerful incentive for renovation firms to comply.

Other commenters observed that prior notification for every covered renovation would be too burdensome for the regulated community and for the agency. Some of these commenters suggested that notifications only be required for renovations involving high-risk methods, housing where a child under age 6 or a pregnant woman resides, or renovations involving multiple rooms in a housing unit.

This final rule does not include a prior notification requirement. EPA disagrees with the notion that there is no way to enforce this regulation without a prior notification requirement. As stated in the preceding discussion on recordkeeping, EPA believes that a review of a renovation firm's records will demonstrate whether or not a renovation firm generally complies with the regulations. In addition, as at least one commenter noted, many renovations require a building permit from the local permitting authority. EPA can work with the local authorities to identify inspection targets. EPA can also follow up on tips and complaints.

EPA agrees with those commenters who believe that prior notification for every project is simply too burdensome for the regulated community and for the agency. If the streamlined, telephone-based system recommended by some of the commenters were implemented, it would reduce the initial burden on the renovation firms. However, EPA would still have to process millions of such notifications annually, and the collective burden on renovation firms and the government would be considerable. Rather than require millions of notifications annually, the great majority of which would never be reviewed, EPA prefers to use other methods for targeting renovation projects for inspections.

An initially attractive option considered by EPA was a prior notification requirement for a subset of covered renovation projects. This option could potentially reduce the notifications received to a manageable level, while preserving the benefits of a prior notification requirement, but EPA was unable to develop appropriate criteria for defining which renovations would require prior notification. EPA considered requiring prior notification for renovations using certain high-risk practices, the practices prohibited by the HUD Lead-Safe Housing Rule and EPA's Lead-Based Paint Activities Regulations. However, EPA ultimately decided, as described in Unit III.E.6 of the preamble, to prohibit most of those practices for covered renovations.

Requiring prior notifications only for renovations in housing where a child under age 6 resides and in child-occupied facilities would not significantly reduce the notifications that would be required. EPA determined that a prior notification requirement tied to project size would not be feasible or effective, because the hazard potential from a renovation job is a combination of the size of the project and the activity being performed.

With regard to the compliance mindset mentioned by some commenters, EPA believes that the recordkeeping requirements are a less burdensome way to achieve the same goal. In fact, a prior notification requirement could lead to EPA targeting for inspection those persons who are most likely to be making an effort to comply with the substantive requirements of the regulation. The person who would not bother to comply with the substantive provisions of this rule would most likely avoid filing a prior notification to EPA before beginning a covered renovation, repair, or painting project. These persons are more likely to be performing renovations in a noncompliant manner than are persons who have complied with a prior notification requirement and told EPA where to find them.

EPA has therefore determined that a prior notification requirement is not an effective or efficient means of facilitating the monitoring of compliance with this regulation. States, territories, and tribes developing their own renovation, repair, and painting programs may come to a different conclusion. These jurisdictions are free to establish prior notification schemes that make sense for their community.

▶ STATE, TERRITORIAL, AND TRIBAL PROGRAMS

Because of the enormous number of renovation activities that occur in this country on an annual basis, EPA welcomes the help of its state, territorial, and tribal partners to ensure that these renovations are performed by trained persons in accordance with this final rule. This final rule establishes, in accordance with TSCA section 404 and EPA's Policy for the Administration of Environmental Programs on Indian Reservations, requirements for the authorization of state, territorial, and tribal renovation, repair, and painting programs. The process for obtaining authorization to operate these programs in lieu of the federal program is the same process used to authorize state, territorial, and tribal lead-based paint activity or pre-renovation education programs found in 40 CFR, Part 745, Subpart Q. Interested states, territories, and Indian tribes may apply for, and receive authorization to, administer and enforce all of the elements of the new subpart E, as amended. States, territories, and tribes may choose to administer and enforce just the existing requirements of subpart E, the pre-renovation education elements, or all of the requirements of the proposed subpart E, as amended.

The 2006 Proposal and the 2007 Supplemental Proposal would not have provided for the authorization of state, territorial, or tribal programs that include only the training, certification, accreditation, and work-practice requirements for renovation, repair, and painting programs and not the pre-renovation education provisions of Subpart E. EPA proposed this approach because the agency believes that the pre-renovation education provisions are an integral part of ensuring that consumers have the information they need to make informed decisions about renovation practices in their homes and other buildings.

In addition, consistent with the proposals, this final rule encourages renovation firms to use the existing pamphlet acknowledgment process to provide owner-occupants of target housing with the opportunity to opt out of the training, certification, and work practice requirements of the rule if they reside in the housing to be renovated, there is no child under age 6 or pregnant woman in residence, the housing does not otherwise meet the definition of COF, and the owner acknowledges that the work practices to be used during the renovation will not necessarily include all of the lead-safe work practices contained in EPA's Renovation, Repair, and Painting Rule.

One state commenter disagreed with EPA's proposed approach and requested that EPA authorize state, territorial, or tribal programs that incorporate only the training, certification, accreditation, and work practices of this final rule because TSCA section 404 allows states to administer and enforce the standards, regulations, or other requirements established under TSCA section 402 or TSCA section 406, or both. EPA agrees with this commenter's reading of TSCA. Therefore, this final rule provides for the authorization of state, territorial, or tribal programs that include either the pre-renovation education requirements of 40 CFR, Part 745, Subpart E, or the training, certification, accreditation, and work practice requirements of this rule, or both.

States, territories, and tribes that wish to administer and enforce the pre-renovation education provisions of Subpart E, as amended, must include both target housing and child-occupied facilities within the scope of their program. Similarly, states, territories, and tribes that are also interested in obtaining authorization to administer and enforce the training, certification, accreditation, work practice, and recordkeeping elements of Subpart E, as amended, must include both target housing and COFs within the scope of their program. States with existing authorized pre-renovation education programs are required to demonstrate that they have modified their programs to include child-occupied facilities. These states must provide this demonstration no later than the first report submitted pursuant to 40 CFR 745.324(h) on or after April 22, 2009.

The authorization process currently codified at 40 CFR, Part 745, Subpart Q, will be used for the purpose of authorizing state, territorial, and tribal renovation, repair, and painting programs. States, territories, and tribes seeking authority for their programs must obtain public input, then submit an application to EPA. Applications must contain a number of items, including a description of the state, territorial, or tribal program, copies of all applicable statutes, regulations, and standards, and a certification by the State Attorney General, Tribal Counsel, or an equivalent official, that the applicable legislation and regulations provide adequate legal authority to administer and enforce the program. The program description must demonstrate that the state, territorial, or tribal program is at least as protective as the federal program. In this case, the federal program consists of the requirements for training, certification, and accreditation and the work practice standards of this final rule.

One commenter suggested that EPA require states with a currently authorized TSCA 402(a) lead-based paint activities program to submit only an amended application for incorporating the TSCA section 402(c)(3) renovation, repair, and painting program requirements since many of the required documents would be the same as those submitted for the original TSCA 402(a) application. Furthermore, the commenter recommended that a letter from the state agency identified in the original 402(a) authorization application with a synopsis detailing how the state proposes to administer and enforce the renovation, repair, and painting program serve as an amended application. EPA has determined that a new application for authorization for the renovation, repair, and painting program is necessary because there may be a different state agency or consortia of agencies implementing and enforcing this program; a long time may have elapsed since most states submitted their TSCA section 402(a) program application; and many of the requirements within the elements of the renovation, repair, and painting program differ from their counterparts in the lead-based paint activities program.

To be eligible for authorization to administer and enforce the training, certification, accreditation, and work practice requirements of this final rule, state, territorial, and tribal renovation programs must contain certain minimum elements (e.g., work practice standards and procedures and requirements for the certification of individuals and/or firms) that are very similar to the existing minimum elements specified in 40 CFR 745.326(a) for lead-based paint activities programs. In order to be authorized, state, territorial, or tribal programs must have procedures and requirements for the accreditation of training programs, which can be as simple as procedures for accepting training provided by an EPA-accredited provider, or a provider accredited by another authorized state, territorial, or tribal program. Procedures and requirements for the certification of renovators are also necessary.

At a minimum, these must include a requirement that certified renovators have taken accredited training, and procedures and requirements for recertification. State, territorial, and tribal programs applying for authorization must also include work practice standards for renovations that ensure that renovations are conducted only by certified renovation firms and the renovations are conducted using work practices at least as protective as those of the federal program. As is the current practice with lead-based paint activities, EPA will not require state, territorial, or tribal programs to certify both firms and individuals that perform renovations. States, territories, and tribes may choose to certify either firms or individuals, so long as the individuals that perform the duties of renovators are required to take accredited training.

Implementation

To provide interested states, territories, and tribes time to develop, or begin developing renovation, repair, and painting programs in accordance with this rule, EPA did not begin to actively implement the federal program until April 22, 2009, at which time EPA began accepting applications for training program accreditation. Several commenters thought 1 year would be adequate for the purpose of allowing states, territories, and tribes to develop their own programs, while others expressed concern that 1 year would not be enough time to get these programs developed and authorized.

Most commenters who expressed an opinion on this topic generally agreed that an implementation delay is necessary. Reasons given in support of a delay were conservation of state financial and administrative resources and the fact that some states have had difficulties in retraining contractors to new state-specific requirements after the contractors had become accustomed to working under the federal program. In contrast, some commenters argued that, in light of the 2010 goal, no delay whatsoever was warranted. This final rule retains the 1 year implementation delay set forth in the 2006 Proposal. EPA determined that this period of time represents an appropriate balance between the need to implement this rule quickly and concerns over potential duplication of effort and costs incurred by the regulated community if EPA begins accrediting training providers and certifying firms in jurisdictions that are also working toward implementing their own programs. States, territories, and tribes may begin the authorization process at any time after the effective date of this final rule, even after the federal program has been implemented in their jurisdiction.

Some commenters were concerned about the effect of this rule on existing state programs. Several commenters asked EPA to expressly state that this rule does not preempt existing state programs and that state programs that are more stringent than the federal program will be eligible for authorization. One commenter noted that the number of houses with lead contaminated paint is disproportionately distributed throughout the U.S. This commenter pointed out that this apparent disparity supports the need for state control of lead programs and for EPA to practice "regulatory restraint." According to this commenter, this "regulatory restraint" will allow states with more severe lead paint problems to impose stricter standards and requirements regarding certification and work practices without imposing unnecessary burdens on states with less severe problems.

This final rule does not preempt existing programs that address renovations. However, to the extent that these programs are less protective than the requirements of this final rule, the requirements of this final rule will apply. To be eligible for authorization, state, territorial, and tribal programs need not exactly duplicate the federal program contained in this final rule, but they must still meet the requirement of TSCA section 404 that they be "at least as protective as" the federal program. It would be difficult for the agency to describe specific requirements that would make a program more or less "protective." EPA will review each program application separately against the protections provided by this final rule.

Several commenters expressed concern regarding the uniformity and consistency of state programs. Some recommended that EPA take states' concerns into account, but guarantee uniformity of state programs by prohibiting states from arbitrarily deviating from program elements. Others noted that if there are uniform regulations for approved training courses for state certification, there should be reciprocity between states since many people work in multiple states.

One commenter suggested that, in an effort to promote consistency, states institute a lead-safety test that renovators must pass prior to receiving permits to conduct work. Several commenters noted that a lack of reciprocity between states and/or duplicative or divergent certification requirements will add an unnecessary burden and level of complexity for renovation and remodeling firms, especially those working in multistate areas. One commenter argued that this could lead to a problem in maintaining certifications similar to the problem the commenter believes exists in maintaining lead-based paint inspector, risk assessor, and other certifications associated with TSCA section 402 abatements. One suggested that EPA should exert control over the right to refuse approval of state programs unless they provide for reciprocity with the federal program and programs of other jurisdictions approved by EPA.

The standard of EPA review for state, territorial, and tribal programs under TSCA section 404 is that they be "at least as protective" as the federal program. In addition, TSCA section 404 (e) reserves the right of states and their political subdivisions to impose requirements that are more stringent than the federal program. EPA interprets this to mean that EPA cannot compel states, territories, and tribes to adopt programs identical to the federal program or to establish reciprocity provisions. However, EPA continues to encourage states, territories, and tribes that may be considering establishing their own renovation programs to keep reciprocity in mind as they move forward.

The benefits to be derived from reciprocity arrangements with the federal program and other authorized jurisdictions include potential cost-savings from reducing duplicative activity and the development of a professional renovation workforce more quickly, thus providing maximum flexibility to state, territorial, or tribal residents. In addition, the agency encourages states, territories, and tribes to consider the use of existing certification and accreditation procedures as they develop their programs. These existing programs need not be limited to lead-based paint. For example, a state may choose to add lead-safe renovation requirements to its existing contractor licensing programs.

▶ EFFECTIVE DATE AND IMPLEMENTATION DATES

This final rule was effective on June 23, 2008, and was implemented according to the following schedule:

1. As of June 23, 2008.
   a. States, territories, and tribes began applying for authorization to administer and enforce their own renovation, repair, and painting programs. EPA began authorizing states, territories, and tribes as soon as it received their complete applications.
   b. No training program may provide, offer, or claim to provide training or refresher training for EPA certification as a renovator or a dust sampling technician without accreditation from EPA under 40 CFR 745.225.
2. As of April 22, 2009. Training programs for renovators or dust sampling technicians began applying for accreditation under 40 CFR 745.225. EPA began accrediting training programs as soon as it received complete applications from training providers. Individuals who wished to become certified renovators or dust sampling technicians began taking accredited training as soon as it was available.

3. As of October 22, 2009. Renovation firms began applying for certification under 40 CFR 745.89. EPA began certifying renovation firms as soon as it received their complete applications.

4. As of April 22, 2010. The rule was fully implemented.

   a. No firm may perform, offer, or claim to perform renovations without certification from EPA under 40 CFR 745.89 in target housing or COFs, unless, in the case of owner-occupied target housing, the firm has obtained a statement signed by the owner that the renovation will occur in the owner's residence, no child under age 6 resides there, the housing is not a child-occupied facility, and the owner acknowledges that the work practices to be used during the renovation will not necessarily include all of the lead-safe work practices contained in EPA's Renovation, Repair, and Painting Rule.

   b. All renovations must be directed by renovators certified in accordance with 40 CFR 745.90(a) and performed by certified renovators or individuals trained in accordance with 40 CFR 5.90(b)(2) in target housing or child-occupied facilities, unless, in the case of owner-occupied target housing, the firm performing the renovation has obtained a statement signed by the owner that the renovation will occur in the owner's residence, no child under age 6 resides there, the housing is not a COF, and the owner acknowledges that the work practices to be used during the renovation will not necessarily include all of the lead-safe work practices contained in EPA's Renovation, Repair, and Painting Rule.

   c. All renovations must be performed in accordance with the work practice standards in 40 CFR 745.85 and the associated recordkeeping requirements in 40 CFR 745.86(b)(6) and (b)(7) in target housing or child-occupied facilities, unless, in the case of owner-occupied target housing, the firm performing the renovation has obtained a statement signed by the owner that the renovation will occur in the owner's residence, no child under age 6 resides there, the housing is not a COF, and the owner acknowledges that the work practices to be used during the renovation will not necessarily include all of the lead-safe work practices contained in EPA's Renovation, Repair, and Painting Rule.

With respect to the new renovation-specific pamphlet and the requirements of the Pre-Renovation Education Rule, as of the effective date of the rule, June 23, 2008, renovators or renovation firms performing renovations in states and Indian tribal areas without an authorized Pre-Renovation Education Rule program may provide owners and occupants with either of the following EPA pamphlets: *Protect Your Family from Lead in Your Home*; or *Renovate Right: Important Lead Hazard Information for Families, Child Care Providers and Schools*. As of December 22, 2008, *Renovate Right: Important Lead Hazard Information for Families, Child Care Providers and Schools* must be used exclusively.

▶ STATUTORY AND EXECUTIVE ORDER REVIEWS

Under Executive Order 12866, entitled *Regulatory Planning and Review* (58 FR 51735, October 4, 1993), it has been determined that this rule is a "significant regulatory action" under section 3(f)(1) of the Executive Order because EPA estimates that it will have an annual effect on the economy of $100 million or more. Accordingly, this action was submitted to the Office of Management and Budget (OMB) for review under Executive Order 12866 and any changes made based on OMB recommendations have been documented in the public docket for this rulemaking as required by section 6(a)(3)(E) of the Executive Order.

In addition, EPA has prepared an analysis of the potential costs and benefits associated with this rulemaking. This analysis is contained in the Economic Analysis, which is available in the docket for this action and is briefly summarized here.

1. *Types of facilities*. This rule applies to an estimated 37.8 million pre-1978 facilities. Of these, approximately 37.7 million facilities are located in target housing, either in rental housing, owner-occupied housing where a child under age 6 resides, or owner-occupied housing where no child under age 6 resides but that otherwise meets the definition of a child-occupied facility. Approximately 100,000 facilities are COFs in pre-1978 public or commercial buildings.
2. *Options evaluated*. EPA considered a variety of options for addressing the risks presented by renovation, repair, and painting actions where lead-based paint is present. The

Economic Analysis analyzed several different options for the scope of the rule, which would limit the coverage of the rule's substantive provisions depending on when the facility was built (such as pre-1960 or pre-1978), and whether or not there are children under the age of 6 or a pregnant woman residing in owner-occupied housing. In some options, coverage of the rule was phased in over time. EPA also considered different options for work practices, such as containment, cleaning, and cleaning verification.

3. *Number of events and individuals affected.* In the first year that all of the rule requirements will be in effect, there will be an estimated 8.4 million renovation, repair, and painting events where lead-safe work practices will be used due to the rule. As a result, there will be approximately 1.4 million children under the age of 6 who will be affected by having their exposure to lead dust minimized due to the rule. There will also be about 5.4 million adults who will be affected. After improved test kits for determining whether a painted surface contains lead-based paint become available (which is assumed in the analysis to occur by the second year of the rule), the number of renovation, repair, and painting events using lead-safe work practices is expected to drop to 4.4 million events per year. No change in the number of exposures avoided due to the rule is expected because the improved test kit will more accurately identify paint without lead, thus reducing the number of events unnecessarily using the required work practices.

4. *Benefits.* The Economic Analysis describes the estimated benefits of the rulemaking in qualitative and quantitative terms. Benefits result from the prevention of adverse health effects attributable to lead exposure. These health effects include impaired cognitive function in children and several illnesses in children and adults. EPA estimated the benefits of avoided incidence of IQ loss due to reduced lead exposure to children under the age of 6. There are not sufficient data at this time to develop dose-response functions for other health effects in children or for pregnant women. The benefits of avoided exposure to adults were not quantified due to uncertainties about the exposure of adults to lead in dust from renovation, repair, and painting activities in these facilities.

The rule is estimated to result in quantified benefits of approximately $700 to $1700 million in the first year. The 50-year annualized benefits provide a measure of the steady-state benefits. The quantified IQ benefits to children are expected to be approximately $700 to $1700 million per year when annualized using a 3% discount rate, and $700 to $1800 million per year when using a 7% discount rate. The estimated benefits for the other scope options range from approximately $300 to $1700 million using a 3% discount rate and from $300 to $1800 million using a 7% discount rate. The benefits from prohibiting certain paint preparation and removal practices in renovations requiring lead-safe work practices under the rule are estimated to be $400 to $900 million per year using a 3% discount rate. There are additional unquantified benefits, including other avoided health effects in children and adults.

5. *Costs.* The Economic Analysis estimates the costs of complying with the rule. Costs may be incurred by contractors who perform renovation, repair, and painting work for compensation; landlords who use their own staff to perform renovation, repair, and painting work in leased buildings; and child-occupied facilities that use their own staff to perform renovation, repair, and painting work.

The rule is estimated to result in a total cost of approximately $800 million in the first year that all of the rule requirements will be in effect. The cost is estimated to drop to approximately $400 million per year in the second year when the improved test kits are assumed to become available. The 50-year annualized costs provide a measure of the steady-state cost. Annualized costs of the rule are estimated to be approximately $400 million per year using either a 3% discount rate or a 7% discount rate. Annualized costs for the other scope options range from approximately $300 million to approximately $700 million per year using a 3% discount rate and $400 to $700 million per year using a 7% discount rate. The cost of prohibiting certain paint preparation and removal practices is estimated to cost less than $10 million per year using either a 3% or a 7% discount rate.

6. *Net benefits.* Net benefits are the difference between benefits and costs. The rule is estimated to result in net benefits of $50 to $1000 million in the first year, based on children's IQ benefits alone. The 50-year annualized net benefits for the rule based on children's benefits are estimated to be $300 to $1300 million per year using either a 3% or a 7% discount

rate. The annualized net benefits for the other scope options range from approximately $50 to $1300 million per year using either a 3% or a 7% discount rate. The net benefits of prohibiting certain paint preparation and removal practices for renovations requiring lead-safe work practices are estimated to be approximately $400 to $900 million per year using either a 3% or a 7% discount rate. There are additional unquantified benefits, including other avoided health effects in children and adults that are not included in the net benefits estimates.

It is important to note that the EPA analysis generates certain results that seem to indicate that more stringent control options yield smaller improvements reducing the risks of elevated blood-lead levels in children than do less stringent control options. For example, the analysis estimates that using only containment of dust and debris generated during an RRP activity yields higher benefits than using all of the rule's work practices (containment, specialized cleaning, and cleaning verification). This is the opposite of what one might expect and of what is observed in the Dust Study for the 10 experiments that used the proposed rule cleaning and containment, since the benefits analysis implies that the combination of rule-style containment with rule-style cleaning and verification would result in more exposure than when such containment is combined with conventional cleaning.

This is inconsistent with the Dust Study, which shows that the largest decreases were observed in the 10 experiments where this final rule's practices of containment, specialized cleaning, and cleaning verification were used. Therefore, the anomalous results are likely to be artifacts of sparse underlying data and modeling assumptions. Although EPA summarizes some of the potential causes of these unexpected results in the Economic Analysis, at this time EPA is unclear as to precisely what is leading to these unexpected results. Because EPA has not determined why the benefits analyses contain anomalous results, EPA has limited confidence in the estimated benefits. EPA does not view the results as being sufficiently robust to represent the difference in magnitude of the benefits across regulatory alternatives. Nevertheless, EPA is confident that there are positive benefits.

► PAPERWORK REDUCTION ACT

The information collection requirements contained in this rule have been submitted for approval to the Office of Management and Budget (OMB) under the Paperwork Reduction Act, 44 U.S.C. 3501 et seq. An Information Collection Request (ICR) document prepared by EPA, an amendment to an existing ICR and referred to as the ICR Final Rule Addendum (EPA ICR No. 1715.10, OMB Control Number 2070-0155), has been placed in the public docket for this rule. The information collection requirements are not enforceable until OMB approves them.

The new information collection activities contained in this rule are designed to assist the agency in meeting the core objectives of TSCA section 402, including ensuring the integrity of accreditation programs for training providers, providing for the certification of renovators, and determining whether work practice standards are being followed. EPA has carefully tailored the recordkeeping requirements so they will permit the agency to achieve statutory objectives without imposing an undue burden on those firms that choose to be involved in renovation, repair, and painting activities.

Burden under the Paperwork Reduction Act means the total time, effort, or financial resources expended by persons to generate, maintain, retain, disclose, or provide information to or for a federal agency. This includes the time needed to review instructions; develop, acquire, install, and utilize technology and systems for the purposes of collecting, validating, and verifying information, processing and maintaining information, and disclosing and providing information; adjust the existing ways to comply with any previously applicable instructions and requirements; train personnel to be able to respond to a collection of information; search data sources; complete and review the collection of information; and transmit or otherwise disclose the information.

Under this rule, the new information collection requirements may affect training providers and firms that perform renovation, repair, or painting for compensation. Although these firms have the option of choosing to engage in the covered activities, once a firm chooses to do so, the information collection activities contained in this rule become mandatory for that firm.

The ICR document provides a detailed presentation of the estimated burden and costs for 3 years of the program. The aggregate burden varies by year due to changes in the number of firms that will seek certification each year. The burden and cost to training providers and firms engaged in renovation, repair, and painting activities is summarized below.

It is estimated that approximately 170 training providers will incur burden and notify EPA (or an authorizing state, tribe, or territory) before and after training courses. The average burden for training provider notifications is estimated at 20 to 100 hours per year, depending on the number of training courses provided. Total training provider burden is estimated to average 9000 hours per year. There are approximately 211,000 firms estimated to become certified to engage in renovation, repair, or painting activities. The average certification burden is estimated to be 3.5 hours per firm in the year a firm is initially certified, and 0.5 hours in years that it is recertified (which occurs every 5 years). Firms must also distribute lead hazard information to the owners and occupants of public or commercial buildings that contain COFs and in target housing containing child-occupied facilities. Finally, firms must keep records of the work they perform; this recordkeeping is estimated to average approximately 5 hours per year per firm. The total burden for these certified firms is estimated to average 1,373,000 hours per year. Total respondent burden during the period covered by the ICR is estimated to average approximately 1,382,000 hours per year.

There are also government costs to administer the program. States, tribes, and territories are allowed, but are under no obligation, to apply for and receive authorization to administer these requirements. EPA will directly administer programs for states, tribes, and territories that do not become authorized. Because the number of states, tribes, and territories that will become authorized is not known, administrative costs are estimated assuming that EPA will administer the program everywhere. To the extent that other government entities become authorized, EPA's administrative costs will be lower.

An agency may not conduct or sponsor, and a person is not required to respond to a collection of information, unless it displays a currently valid OMB control number. The OMB control numbers for EPA's regulations codified in Chapter 40 of the CFR, after appearing in the preamble of the final rule, are listed in 40 CFR, Part 9, displayed either by publication in the *Federal Register* or by other appropriate means, such as on the related collection instrument or form, if applicable. When this ICR is approved by OMB, the agency will publish a technical amendment to 40 CFR, Part 9, in the *Federal Register* to display the OMB control number for the approved information collection requirements contained in this final rule.

▶ REGULATORY FLEXIBILITY ACT

The Regulatory Flexibility Act (RFA) generally requires that an agency prepare a regulatory flexibility analysis of any rule subject to notice and comment rulemaking requirements under the Administrative Procedure Act or any other statute unless the agency certifies that the rule will not have a significant economic impact on a substantial number of small entities. Small entities include small businesses, small organizations, and small governmental jurisdictions.

For purposes of assessing the impacts of this rule on small entities, small entity is defined in accordance with section 601 of the RFA as: (1) a small business as defined by the Small Business Administration's (SBA) regulations at 13 CFR 121.201; (2) a small governmental jurisdiction that is a government of a city, county, town, school district, or special district with a population of less than 50,000; and (3) a small organization that is any not-for-profit enterprise which is independently owned and operated and is not dominant in its field.

Pursuant to section 603 of the RFA, EPA prepared an initial regulatory flexibility analysis (IRFA) for the proposed rule and convened a Small Business Advocacy Review Panel to obtain advice and recommendations of representatives of the regulated small entities. A summary of the IRFA, a description of the Panel process, and a summary of the Panel's recommendations can be found in Unit VIII.C. of the preamble to the 2006 Proposal. A detailed discussion of the Panel's advice and recommendations is found in the Panel Report.

As required by section 604 of the RFA, final regulatory flexibility analysis (FRFA) was also prepared for this final rule. The FRFA addresses the issues raised by public comments on the IRFA, which was part of the proposal of this rule. The FRFA is available for review in the docket and is summarized next.

1. *Legal basis and objectives for the rule.* As discussed in Unit II.A of this preamble, TSCA section 402(c)(2) directs EPA to study the extent to which persons engaged in renovation, repair, and painting activities are exposed to lead or create lead-based paint hazards regularly or occasionally. After concluding this study, TSCA section 402(c)(3) further directs EPA to revise its Lead-Based Paint Activities Regulations under TSCA section 402(a) to apply to renovation or remodeling activities that create lead-based paint hazards. Because EPA's study found that activities commonly performed during renovation and remodeling create lead-based paint hazards, EPA is revising the TSCA section 402(a) regulatory scheme to apply to individuals and firms engaged in renovation, repair, and painting activities. In so doing, EPA has also taken into consideration the environmental, economic, and social impact of this final rule as provided in TSCA section 2(c). The primary objective of the rule is to minimize exposure to lead-based paint hazards created during renovation, repair, and painting activities in housing where children under age 6 reside, in housing where a pregnant woman resides, and in housing or other buildings frequented by children under age 6.

2. *Potentially affected small entities.* Small entities include small businesses, small organizations, and small governmental jurisdictions. The small entities that are potentially directly regulated by this rule include small businesses (including contractors and property owners and managers); small nonprofits (certain daycare centers and private schools); and small governments (school districts).

   In determining the number of small businesses affected by the rule, the agency applied U.S. Economic Census data to the SBA's definition of small business. However, applying the U.S. Economic Census data requires either under- or overestimating the number of small businesses affected by the rule. For example, for many construction establishments, the SBA defines small businesses as having revenues of less than $13 million.

   With respect to those establishments, the U.S. Economic Census data groups all establishments with revenues of $10 million or more into one revenue bracket. On the one hand, using data for the entire industry would overestimate the number of small businesses affected by the rule and would defeat the purpose of estimating impacts on small business. It would also underestimate the rule's impact on small businesses because the impacts would be calculated using the revenues of large businesses in addition to small businesses. On the other hand, applying the closest, albeit lower, revenue bracket would underestimate the number of small businesses affected by the rule while at the same time overestimating the impacts.

Similar issues arose in estimating the fraction of property owners and managers that are small businesses. EPA has concluded that a substantial number of small businesses will be affected by the rule. Consequently, EPA has chosen to be more conservative in estimating the cost impacts of the rule by using the closest, albeit lower, revenue bracket for which Census data is available. For other sectors (nonprofits operating daycare centers or private schools), EPA assumed that all affected firms are small, which may overestimate the number of small entities affected by the rule.

The vast majority of entities in the industries affected by this rule are small. Using EPA's estimates, the renovation, repair, and painting program will affect an average of approximately 189,000 small entities.

3. *Potential economic impacts on small entities.* EPA evaluated two factors in its analysis of the rule's requirements on small entities, the number of firms that would experience the impact, and the size of the impact. Average annual compliance costs as a percentage of average annual revenues were used to assess the potential average impacts of the rule on small businesses and small governments. This ratio is a good measure of entities' ability to afford the costs attributable to a regulatory requirement, because comparing compliance costs to revenues provides a reasonable indication of the magnitude of the regulatory burden relative to a commonly available measure of economic activity. Where regulatory costs represent a small fraction of a typical entity's revenues, the financial impacts of the regulation on such entities may be considered as not significant. For nonprofit organizations, impacts were measured by comparing rule costs to annual expenditures. When expenditure data were not available, however, revenue information was used as a proxy for expenditures. It is

appropriate to calculate the impact ratios using annualized costs, because these costs are more representative of the continuing costs entities face to comply with the rule.

EPA estimates that there are an average of 189,000 small entities that would be affected by the renovation, repair, and painting activities program. Of these, there are an estimated 165,000 small businesses with an average impact of 0.7%, 17,000 small nonprofits with an average impact of 0.1%, and 6000 small governments with an average impact of 0.004%. These estimates are based on an average cost of approximately $35 per renovation.

4. *Relevant federal rules.* The requirements in this rulemaking will fit within an existing framework of other federal regulations that address lead-based paint. The Pre-Renovation Education Rule, discussed in Unit II.A.2 of the preamble, requires renovators to distribute a lead hazard information pamphlet to owners and occupants before conducting a renovation in target housing. This rule has been carefully crafted to harmonize with the existing pre-renovation education requirements.

Disposal of waste from renovation projects that would be regulated by this rule is covered by the Resource Conservation and Recovery Act (RCRA) regulations for solid waste. This rule does not contain specific requirements for the disposal of waste from renovations.

HUD has extensive regulations that address the conduct of interim controls, as well as other lead-based paint activities, in federally assisted housing. Some of HUD's interim controls are regulated under this rule as renovations, depending on whether the particular interim control measure disturbs more than the threshold amount of paint. In most cases, the HUD regulations are comparable to, or more stringent than this rule. In general, persons performing HUD-regulated interim controls must have taken a course in lead-safe work practices, which is also a requirement of this rule. However, this rule does not require dust clearance testing, a process required by HUD after interim control activities that disturb more than a minimal amount of lead-based paint.

Finally, OSHA's Lead Exposure in Construction standard covers potential worker exposures to lead during many construction activities, including renovation, repair, and painting activities. Although this standard may cover many of the same projects as this final rule, the requirements themselves do not overlap. The OSHA rule addresses the protection of the worker; this EPA rule principally addresses the protection of the building occupants, particularly children under age 6 and pregnant women.

5. *Skills needed for compliance.* This rule establishes requirements for training renovators, other renovation workers, and dust sampling technicians; certifying renovators, dust sampling technicians, and entities engaged in renovation, repair, and painting activities; accrediting providers of renovation and dust sampling technician training; and for renovation work practices. Renovators and dust sampling technicians would have to take a course to learn the proper techniques for accomplishing the tasks they will perform during renovations. These courses are intended to provide them with the information they would need to comply with the rule based on the skills they already have. Renovators would then provide on-the-job training in work practices to any other renovation workers used on a particular renovation. They would also need to document the work they have done during renovations. This does not require any special skills. Renovation firms would be required to apply for certification to perform renovations; this process does not require any special skills other than the ability to complete the application. Training providers must be knowledgeable about delivering technical training. Training providers would be required to apply for accreditation to offer renovator and dust sampling technician courses. They would also be required to provide prior notification of such courses and provide information on the students trained after each such course. Completing the accreditation application and providing the required notification information does not require any special skills.

6. *Small Business Advocacy Review Panel.* Since the earliest stages of planning for this regulation under section 402(c)(3) of TSCA, EPA has been concerned with potential small entity impacts. EPA conducted outreach to small entities, and, in 1999, convened a Small Business Advocacy Review (SBAR) Panel to obtain advice and recommendations of representatives of the small entities that would potentially be subject to

this regulation's requirements. At that time, EPA was planning an initial regulation that would apply to renovations in target housing, with requirements for public and commercial building renovations, including COF renovations, to follow at a later date. The small entity representatives (SERs) chosen for consultation reflect that initial emphasis. They included maintenance and renovation contractors, painting and decorating contractors, multifamily housing owners and operators, training providers/consultants, and representatives from several national contractor associations, the National Multi Housing Council, and the National Association of Home Builders. After considering the existing Lead-Based Paint Activities Regulations, and taking into account preliminary stakeholder feedback, EPA identified these eight key elements of a potential renovation and remodeling regulation for the SBAR Panel's consideration.

- Applicability and scope
- Firm certification
- Individual training and certification
- Accreditation of training courses
- Work practice standards
- Prohibited practices
- Exterior clearance
- Interior clearance

EPA also developed several options for each of these key elements. Although the scope and applicability options specifically presented to the SBAR Panel covered only target housing, background information presented to the SERs and to the SBAR Panel members shows that EPA was also considering a regulation covering child-occupied facilities. The 2007 Supplemental Proposal extended the potentially regulated universe to include COFs. When the 2007 Supplemental Proposal was issued, EPA conducted a targeted mailing campaign to specifically solicit input on the rule from child-occupied facilities, such as childcare providers and kindergartens, in public or commercial buildings. More information on the SBAR Panel, its recommendations, and how EPA implemented them in the development of the program, is provided in Unit VIII.C.6 of the preamble to the 2006 Proposal.

Alternatives Considered

The following is a discussion of significant alternatives to the rule, originated by EPA or by commenters, that could affect the economic impacts of the rule on small entities. These alternatives would have applied to both small and large entities, but, given the large number of small entities in the industry, these alternatives would primarily affect small entities. For the reasons described below, these alternatives are not consistent with the objectives of the rule. EPA considered a number of options for the scope and applicability of the rule: include all pre-1978 housing, all pre-1978 rental housing, all pre-1960 housing, and all pre-1960 rental housing. Although the scope and applicability options specifically presented to the SBAR Panel covered only target housing, background information presented to the SERs and to the SBAR Panel members shows that EPA was also considering a regulation covering COFs.

The SBAR Panel recommended that EPA request public comment in the proposal on the option of limiting the housing stock affected by the rule to that constructed prior to 1960, as well as the option of covering all pre-1978 housing and other options that may help to reduce costs while achieving the protection of public health. EPA asked for comment in the proposed rule on alternative scope options, including an option limited to buildings constructed prior to 1960. After considering the public comments, EPA determined that limiting the rule to exclude buildings constructed on or after 1960 was not consistent with the stated objectives of the rule, in part because this would not protect children under the age of 6 and pregnant women.

EPA proposed a staged approach that would initially address renovations in pre-1960 target housing and child-occupied facilities, or where a child had an increased blood-lead level. EPA requested comment about whether to delay implementation for post-1960 target housing and COFs for 1 year. Most commenters objected to the phased implementation, expressing concerns about adding complexity to implementation and about potential exposures to children in buildings built between 1960 and 1978 during the first year. After reviewing the comments, EPA determined the reduced burdens of a staged approach did not outweigh the complexity that it added to implementation.

Exclude Categories of Contractors or Renovation Activities

EPA requested comment on whether to exclude any categories of specialty contractors and whether certain renovation activities should be specifically included or excluded. In response, no commenter offered any data to show that any category of contractor or type of renovation activity should be exempt because they do not create lead-based paint hazards. All of the renovation activities in the Dust Study and the other studies in the record for the rule created lead-based paint hazards. EPA determined that it had no basis on which to exempt any category of contractor or type of renovation. However, some small jobs will be exempt from the requirements of the rule under the minor maintenance exception.

The current abatement regulations in 40 CFR, Part 745, Subpart L, prohibit the following work practices during abatement projects: open-flame burning or torching, machine sanding or grinding, abrasive blasting or sandblasting, dry scraping of large areas, and operating a heat gun in excess of 1100°F. EPA presented four options to the SBAR Panel on this topic: prohibit these practices during renovations; allow dry scraping and exterior flame-burning or torching; allow dry scraping and interior and exterior flame-burning or torching; or allow all of these practices. The SBAR Panel recognized industry concerns over the feasibility of prohibiting these practices, especially when no cost-effective alternatives exist. The SBAR Panel was also concerned about the potential risks associated with these practices, but noted that reasonable training, performance, containment, and cleanup requirements may adequately address these risks.

EPA followed the SBAR Panel's recommendation and requested public comment on the cost, benefit, and feasibility of prohibiting certain work practices. In response to its request for comment in the proposed rule, the agency received information on techniques including benign strippers, steam stripping, closed planing with vacuums, infrared removal, and chemical stripping. Therefore, EPA believes that there are cost-effective alternatives to these prohibited or restricted practices. In addition, the Dust Study ("Characterization of Dust Lead Levels after Renovation, Repair, and Painting Activities") found that most practices prohibited or restricted under EPA's Lead-Based Paint Activities Regulations produce large quantities of lead dust, and that the use of the proposed work practices was not effective at containing or removing dust-lead hazards from the work area.

EPA has concluded that these practices should be prohibited or restricted during renovation, repair, and painting activities that disturb lead-based paint because the work practices in the rule are not effective at containing the spread of leaded dust when these practices are used, or at cleaning up lead-based paint hazards created by these practices. Thus, the work practices are not effective at minimizing exposure to lead-based paint hazards created during renovation activities when these activities are used.

The proposed rule required the use of a HEPA vacuum as part of the work practice standards for renovation activities. One commenter stated that EPA did not have sufficient evidence showing that HEPA vacuums are significantly better at removing lead dust than non-HEPA vacuums. EPA has determined that the weight of the evidence provided by the studies it reviewed demonstrates that the HEPA vacuums consistently removed significant quantities of lead-based paint dust and reduced lead loadings to lower levels then did other vacuums. While there may be some vacuum cleaners that are as effective as HEPA vacuums, EPA has not been able to define quantitatively the specific attributes of those vacuums. That is, EPA is not able to identify what criteria should be used to identify vacuums that are equivalent to HEPA vacuums in performance. Thus, EPA does not believe that it can identify in the final rule what types of vacuums can be used as substitutes for HEPA vacuums. Therefore, EPA has not adopted this alternative.

EPA requested comment on whether cleaning verification is necessary given the cleaning required by the rule. Some commenters contended that a visual inspection following cleaning after a renovation is sufficient to ensure the lead-based paint dust generated by a renovation has been sufficiently cleaned up. EPA disagrees with those commenters who requested that the work practices in the final rule not include any verification beyond visual inspection. The weight of the evidence clearly demonstrates that visual inspection following cleaning after a renovation is insufficient at detecting dust-lead hazards, even at levels significantly above the regulatory hazard standards. Further, EPA disagrees with the implication that easily visible paint chips and splinters are necessarily the primary materials generated during a renovation. EPA studies, including the Dust Study, show that renovation activities generate dust as well as chips and splinters. Therefore, EPA has not adopted this alternative.

A commenter requested that the plumbing-heating-cooling industry be exempted from the rule, claiming that the rule is impractical for the industry. The commenter did not provide any supporting data as to why the rule is impractical for the plumbing-heating-cooling industry, or any data indicating that renovations conducted by plumbing, heating, or cooling contractors do not create lead hazards. By contrast, the Dust Study indicated that cutting open drywall (an activity often performed by plumbing, heating, and cooling contractors) can create a lead hazard. Therefore, EPA believes that plumbing, heating, and cooling contractors who disturb more than an exempt amount of lead-based paint can create lead hazards. EPA does not believe that there is a factual basis for exempting this, or any other, industry from the rule.

Another commenter stated that EPA's proposed rule gave little deference to HUD's rules, and thus is inconsistent with the Regulatory Flexibility Act's requirements to fit new rules within the framework of existing federal regulations. The commenter stated that EPA's rule needed to give greater deference to the framework established in HUD's rules (especially HUD's requirements for independent clearance examinations and its prohibition of dangerous work practices), and to clearly explain how the Renovation, Repair, and Painting Rule will interface with HUD's rules to avoid confusion.

Regarding HUD's requirements for independent clearance examinations, EPA's final rule clarifies that dust clearance sampling is allowed in lieu of post-renovation cleaning verification in cases where another federal, state, territorial, tribal, or local regulation requires dust clearance testing and requires the renovation firm to clean the work area until it passes clearance. This would apply to HUD-regulated renovations. Regarding the prohibition of dangerous work practices, EPA's final rule prohibits the use of the following work practices during regulated renovations: open flame burning or torching of lead-based paint; the use of machines that remove lead-based paint through high-speed operation such as sanding, grinding, power planing, needle gun, abrasive blasting, or sandblasting unless such machines are used with HEPA exhaust control; and operating a heat gun above 1100°F. EPA believes that the provisions in the final rule provide an appropriate measure of consistency with other regulatory programs (including HUD's), and will cause minimal disruption for renovation firms.

One commenter contended that EPA said that "none of the housing authorities identified in Section 8.2.1 as operating public housing that does not receive HUD funding qualifies as a small government under the Regulatory Flexibility Act." According to the commenter, public housing authorities are government entities, and hundreds of them are located in and are part of communities with a population of less than 50,000.

EPA's small entity analysis was not claiming that no small governments operate housing authorities, but that they would not be significantly impacted by the rule. EPA's reasoning was as follows:

- The only public housing authorities that EPA could identify that do not receive HUD funds are operated by Massachusetts, New York, Hawaii, Connecticut, and New York City.
- Massachusetts, New York, Hawaii, Connecticut, and New York City have populations over 50,000 and thus do not qualify as small governments.
- To the best of EPA's knowledge, governments with populations under 50,000 that operate public housing authorities all receive HUD funds.
- Public housing that receives funding from HUD already must comply with HUD regulations regarding lead paint and so are not likely to incur significant additional costs due to this rule.

The commenter has offered no factual information to dispute this reasoning. Therefore, the Agency believes its conclusions regarding public housing authorities operated by small governments were appropriate.

A commenter stated that the proposed rule will have a significant impact on small businesses, and that EPA's own economic analysis of this rule finds that residential property managers and lessors of residential real estate will bear the largest share of costs in association with the rule. EPA disagrees with the commenter's claim that residential property managers and lessors of residential real estate will bear the largest share of costs in association with the rule.

EPA analyzed small business impacts by estimating the average cost impact ratio for each industry, calculated as the average annual compliance cost as a percentage of average annual revenues. The average cost impact ratio for lessors of real estate is below the average cost impact ratio for all small businesses under the rule. And while the average cost impact ratio for residential property managers is above the average cost impact for all small businesses under the rule, small residential property managers make up approximately 3% of the small entities impacted by the rule. Therefore, it is not accurate to claim that residential property managers and lessors of residential real estate will bear the largest share of costs in association with the rule.

Another commenter stated that given the lack of evidence showing that HEPA vacuums are significantly better at removing lead dust from floors, and because HEPA vacuums are significantly more costly than non-HEPA units, EPA should modify its proposed rule to allow cleanup with either a HEPA or non-HEPA vacuum. According to the commenter, doing so would reduce the cost to small entities in the renovation and lead mitigation businesses without compromising the level of lead dust clearance achieved by the standard.

EPA disagrees that it should modify its proposed rule to allow cleanup with a non-HEPA vacuum. EPA has determined that the weight of the evidence provided by various studies demonstrates that the HEPA vacuums consistently removed significant quantities of lead-based paint dust and reduced lead loadings to lower levels then did other vacuums. While there may be some vacuums that are as effective as HEPA vacuums, EPA has not been able to define quantitatively the specific attributes of those vacuums. That is, EPA is not able to identify what criteria should be used to identify vacuums that are equivalent to HEPA vacuums in performance. Thus, EPA does not believe that it can identify what types of vacuums can be used as substitutes for HEPA vacuums. EPA also notes that non-HEPA vacuums that perform as well as HEPA vacuums may not be less expensive than HEPA vacuums. For these reasons, EPA has determined that modifying its proposed rule to allow cleanup with non-HEPA vacuums would compromise the level of lead dust clearance achieved by the standard, and might not result in meaningful cost reductions.

As required by Section 212 of SBREFA, EPA is also preparing a Small Entity Compliance Guide to help small entities comply with this rule. Before the date that this rule's requirements take effect for training providers, renovation firms, and renovators, the guide will be available on EPA's website at *www.epa.gov/lead* or from the National Lead Information Center by calling 800-424-LEAD (5323).

▶ UNFUNDED MANDATES REFORM ACT

Title II of the Unfunded Mandates Reform Act of 1995 (UMRA), Public Law 104-4, establishes requirements for federal agencies to assess the effects of their regulatory actions on state, local, and tribal governments and the private sector. Under section 202 of the UMRA, EPA generally must prepare a written statement, including a cost-benefit analysis, for proposed and final rules with "federal mandates" that may result in expenditures to state, local, and tribal governments, in the aggregate, or to the private sector, of $100 million or more in any 1 year. Before promulgating an EPA rule for which a written statement is needed, section 205 of the UMRA generally requires EPA to identify and consider a reasonable number of regulatory alternatives and adopt the least costly, most cost-effective, or least burdensome alternative that achieves the objectives of the rule. The provisions of section 205 do not apply when they are inconsistent with applicable law.

Moreover, section 205 allows EPA to adopt an alternative other than the least costly, most cost-effective, or least burdensome alternative if the administrator publishes with the final rule an explanation why that alternative was not adopted. Before EPA establishes any regulatory requirements that may significantly or uniquely affect small governments, including tribal governments, it must have developed under section 203 of UMRA a small government agency plan. The plan must provide for notifying potentially affected small governments, enabling officials of affected small governments to have meaningful and timely input in the development of EPA regulatory proposals with significant federal intergovernmental mandates, and informing, educating, and advising small governments on compliance with the regulatory requirements.

Under UMRA Title II, EPA has determined that this rule contains a federal mandate that may result in expenditures that exceed the inflation-adjusted UMRA threshold of $100 million by the private sector in any 1 year, but it will not result in such expenditures by state, local, and tribal governments in the aggregate. Accordingly, EPA has prepared a written statement under section 202 of UMRA which has been placed in the public docket for this rulemaking and is summarized here.

Authorizing Legislation

This rule is issued under the authority of TSCA sections 402(c)(3), 404, 406, and 407, 15 U.S.C. 2682(c)(3), 2684, 2686, and 2687.

Cost–Benefit Analysis

EPA has prepared an analysis of the costs and benefits associated with this rulemaking, a copy of which is available in the docket for this rulemaking. The Economic Analysis presents the costs of the rule as well as various regulatory options and is summarized in Unit III.A of the preamble. EPA has estimated that the total annualized costs of this rulemaking are approximately $400 million per year using either a 3% or a 7% discount rate, and that benefits are approximately $700 to $1700 million per year using a 3% discount rate and $700 to $1800 million per year using a 7% discount.

State, Local, and Tribal Government

EPA has sought input from state, local and tribal government representatives throughout the development of the renovation, repair, and painting program. EPA's experience in administering the existing lead-based paint activities program under TSCA section 402(a) suggests that these governments will play a critical role in the successful implementation of a national program to reduce exposures to lead-based paint hazards associated with renovation, repair, and painting activities. Consequently, as discussed in Unit III.C.2 of the preamble to the 2006 Proposal, the agency has met with state, local, and tribal government officials on numerous occasions to discuss renovation issues.

Least Burdensome Option

EPA considered a wide variety of options for addressing the risks presented by renovation activities where lead-based paint is present. As part of the development of the renovation, repair, and painting program, EPA has considered different options for the scope of the rule, various combinations of training and certification requirements for individuals who perform renovations, various combinations of work practice requirements, and various methods for ensuring that no lead-based paint hazards are left behind by persons performing renovations. The Economic Analysis analyzed several different options for the scope of the rule.

Additional information on the options considered is available in Unit VIII.C.6 of the preamble for the 2006 Proposal, and in the Economic Analysis. EPA has determined that the preferred option is the least burdensome option available that achieves the primary objective of this rule, which is to minimize exposure to lead-based paint hazards created during renovation, repair, and painting activities in housing where children under age 6 reside, where a pregnant woman resides, and in housing or other buildings frequented by children under age 6.

This rule does not contain a significant federal intergovernmental mandate as described by Section 203 of UMRA. Based on the definition of "small government jurisdiction" in RFA section 601, no state governments can be considered small. Small territorial or tribal governments may apply for authorization to administer and enforce this program, which would entail costs, but these small jurisdictions are under no obligation to do so.

EPA has determined that this rule contains no regulatory requirements that might significantly or uniquely affect small governments. Small governments operate schools that are child-occupied facilities. EPA generally measures a significant impact under UMRA as being expenditures, in the aggregate, of more than 1% of small government revenues in any 1 year. As explained in Unit III.C.3, the rule is expected to result in small government impacts well under 1% of revenues. So EPA has determined that the rule does not significantly affect small governments. Nor does the rule uniquely affect small governments, as the rule is not aimed at small governments, does not primarily affect small governments, and does not impose a different burden on small governments than on other entities that operate child-occupied facilities.

▶ FEDERALISM

Pursuant to Executive Order 13132, entitled *Federalism* (64 FR 43255, August 10, 1999), EPA has determined that this rule does not have "federalism implications," because it will not have substantial direct effects on the states, on the relationship between the national government and the states, or on the distribution of power and responsibilities among the various levels of government, as specified in Executive Order 13132. Thus, Executive Order 13132 does not apply to this rule. States would be able to apply for, and receive authorization to administer these requirements, but would be under no obligation to do so. In the absence of a state authorization, EPA will administer these requirements. Nevertheless, in the spirit of the objectives of this Executive Order, and consistent with EPA policy to promote communications between the agency and state and local governments, EPA has consulted with representatives of state and local governments in developing the renovation, repair, and painting program. These consultations are as described in the preamble to the 2006 Proposal.

As required by Executive Order 13175, entitled "Consultation and Coordination with Indian Tribal Governments" (59 FR 22951, November 9, 2000), EPA has determined that this rule does not have tribal implications because it will not have substantial direct effects on tribal governments, on the relationship between the federal government and the Indian tribes, or on the distribution of power and responsibilities between the federal government and Indian tribes, as specified in the Order. Tribes would be able to apply for, and receive authorization to administer these requirements on tribal lands, but tribes would be under no obligation to do so. In the absence of a tribal authorization, EPA will administer these requirements. While tribes may operate COFs covered by the rule such as kindergartens, pre-kindergartens, and daycare facilities, EPA has determined that this rule would not have substantial direct effects on the tribal governments that operate these facilities.

Thus, Executive Order 13175 does not apply to this rule. Although Executive Order 13175 does not apply to this rule, EPA consulted with tribal officials and others by discussing potential renovation regulatory options for the renovation, repair, and painting program at several national lead program meetings hosted by EPA and other interested federal agencies.

▶ CHILDREN'S HEALTH PROTECTION

Executive Order 13045, entitled "Protection of Children from Environmental Health Risks and Safety Risks" (62 FR 19885, April 23, 1997), applies to this rule because it is an "economically significant regulatory action" as defined by Executive Order 12866, and because the environmental health or safety risk addressed by this action may have a disproportionate effect on children. Accordingly, EPA has evaluated the environmental health or safety effects of renovation, repair, and painting projects on children. Various aspects of this evaluation are discussed in the preamble to the 2006 Proposal.

The primary purpose of this rule is to minimize exposure to lead-based paint hazards created during renovation, repair, and painting activities in housing where children under age 6 reside and in housing or other buildings frequented by children under age 6. In the absence of this regulation, adequate work practices are not likely to be employed during renovation, repair, and painting activities. EPA's analysis indicates that there will be approximately 1.4 million children under age 6 affected by the rule. These children are projected to receive considerable benefits due to this regulation.

▶ ENERGY EFFECTS

This rule is not a "significant energy action" as defined in Executive Order 13211, entitled "Actions Concerning Regulations That Significantly Affect Energy Supply, Distribution, or Use" (66 FR 28355, May 22, 2001), because it is not likely to have any adverse effect on the supply, distribution, or use of energy.

Technology Standards

Section 12(d) of the National Technology Transfer and Advancement Act of 1995 (NTTAA), Public Law No. 104–113, 12(d) (15 U.S.C. 272 note), directs EPA to use voluntary consensus standards in its regulatory activities unless to do so would be inconsistent with applicable law or otherwise impractical. Voluntary consensus standards are technical standards (e.g., materials specifications, test methods, sampling procedures, and business practices) that are developed or adopted by voluntary consensus standards bodies.

The NTTAA directs EPA to provide Congress, through OMB, explanations when the agency decides not to use available and applicable voluntary consensus standards. In the 2006 Proposal, EPA proposed to adopt a number of work practice requirements that could be considered technical standards for performing renovation projects in residences that contain lead-based paint. As discussed in Unit VIII.I of the 2006 Proposal, EPA identified two potentially applicable voluntary consensus standards (Ref. 3 at 1626). ASTM International (formerly the American Society for Testing and Materials) has developed two potentially applicable documents: "Standard Practice for Clearance Examinations Following Lead Hazard Reduction Activities in Single-Family Dwellings and Child-Occupied Facilities," and "Standard Guide for Evaluation, Management, and Control of Lead Hazards in Facilities." With respect to the first document, EPA did not propose to require traditional clearance examinations, including dust sampling, following renovation projects. However, EPA did propose to require that a visual inspection for dust, debris, and residue be conducted after cleaning and before post-renovation cleaning verification is performed.

The first ASTM document does contain information on conducting a visual inspection before collecting dust clearance samples. The second ASTM document is a comprehensive guide to identifying and controlling lead-based paint hazards. Some of the information in this document is relevant to the work practices required by the rule. Each of these ASTM documents represents state-of-the-art knowledge regarding the performance of these particular aspects of lead-based paint hazard evaluation and control practices, and EPA continues to recommend the use of these documents where appropriate. However, because each of these documents is extremely detailed and encompasses many circumstances beyond the scope of this rulemaking, EPA determined that it would be impractical to incorporate these voluntary consensus standards into the rule.

In addition, this final rule contains performance standards and a process for recognizing test kits that may be used by certified renovators to determine whether components to be affected by a renovation contain lead-based paint. EPA will recognize those kits that meet certain performance standards for limited false positives and negatives. EPA will also recognize only those kits that have been properly validated by a laboratory independent of the kit manufacturer. For most kits, this will mean participating in EPA's Environmental Technology Verification (ETV) program. With stakeholder input, EPA is adapting a voluntary consensus standard, ASTM's "Standard Practice for Evaluating the Performance Characteristics of Qualitative Chemical Spot Test Kits for Lead in Paint," for use as a testing protocol to determine whether a particular kit has met the performance standards established in this final rule.

▶ ENVIRONMENTAL JUSTICE

Executive Order 12898, entitled "Federal Actions to Address Environmental Justice in Minority Populations and Low-Income Populations" (59 FR 7629, February 16, 1994), establishes federal executive policy on environmental justice. Its main provision directs federal agencies, to the greatest extent practicable and permitted by law, to make environmental justice part of their mission by identifying and addressing, as appropriate, disproportionately high and adverse human health or environmental effects of their programs, policies, and activities on minority populations and low-income populations in the United States.

EPA has assessed the potential impact of this rule on minority and low-income populations. The results of this assessment are presented in the Economic Analysis, which is available in the public docket for this rulemaking. As a result of this assessment, the agency has determined that this final rule will not have disproportionately high and adverse human health or environmental effects on minority or low-income populations because it increases the level of environmental protection for all affected populations without having any disproportionately high and adverse human health or environmental effects on any population, including any minority or low-income population.

▶ CONCLUSION

We have reached the end of the line. You should now have a firm grasp of the key elements of working with lead hazards. Never overlook the rules, regulations, and laws that must be followed when working with lead-containing materials. Remember to check your local regulations. Do not permit yourself to be found in a bad situation due to lack of information. The only things missing here are the illustrations mentioned at the beginning of this chapter. They are shown in Figures 9.1 through 9.5.

Multifamily Housing: Component Type Report

Address/Unit No. _Fenway Gardens Housing Complex/Unit #9_

Date _August 16, 2010_　　XRF Serial No. _RS-1967_

Inspector Name _Mo Smith_　　　　　　　　Signature

Description	Total Number	POSITIVE		INCONCLUSIVE				NEGATIVE		Final Class.
		Number	Percent	Low Number	Low Percent	High Number	High Percent	Number	Percent	
Wood Shelves	83	4	4.8	5	6.0	9	.8	65	78.3	NEG
Wood Doors	110	40	36.4	12	10.9	8	7.3	50	45.5	POS
~~Wood Door Casings~~	~~34~~	~~6~~	~~17.6~~	~~5~~	~~14.7~~	~~5~~	~~14.7~~	~~18~~	~~52.9~~	
Wood Hall Cabinets	60	5	8.3	8	13.3	12	20.0	35	58.3	POS
Wood Window Stools	110	60	54.5	30	27.3	10	9.1	10	9.1	POS
Wood Window Casings	63	0	0.0	0	0.0	0.0	0.0	63	100	NEG
Plaster Walls	110	0	0.0	10	9.1	9	8.2	91	82.7	NEG
Concrete Support Columns	40	40	100	0	0.0	0	0.0	0	0.0	POS
Concrete Ceiling Beams	40	40	100	0	0.0	0	0.0	0	0.0	POS
Metal Baseboards	45	0	0.0	0	0.0	0	0.0	45	100	NEG
Metal Gutters	50	20	40.0	8	16.0	2	4.0	20	40.0	POS
Brick Stairway	50	10	20.0	4	8.0	6	12.0	30	60.0	POS
Metal Radiators*	55	0	0.0	11	20.0	13	23.6	31	56.4	POS
Wood Door Casings	40	12	30.0	5	12.5	5	12.5	18	45.0	POS
* Metal Radiators:										
(retest of high inconclusives)	13	9	69.2					4	30.7	POS

1997 Revision　　　　　　　　Completed Form 7.6

* **Lower Boundary:** _0.41 mg/cm²_ **Upper Boundary:** _1.39 mg/cm²_ **Midpoint:** _0.90 mg/cm²_

Figure 9.1 Sample of a completed multifamily housing component type report.

		Yes	No
1	Did the report clearly explain the entire testing program and include an executive summary in narrative form?		
2	Did the report provide an itemized list of similar building components (testing combinations) and the percentage of each component that tested positive, negative, and inconclusive? (Percentages are not applicable for single-family dwellings.)		
3	Did the report include test results for the common areas and building exteriors as well as the interior of the dwelling units?		
4	Were all painted surfaces known to exist in dwelling units, common areas, and building exteriors included in itemized list of components tested?		
5	If confirmation testing (laboratory testing) was necessary, did the testing or inspection firm amend the final report and revise the list of surfaces that tested positive, negative, and inconclusive?		
6	Was the unit selection process performed randomly?		
7	Is the name of the XRF manufacturer and the model, serial numbers of the XRF that was used in each unit recorded in the report?		
8	Did the report record the XRF calibration checks for each day that testing was performed?		
9	Did the calibration checks indicate that the instrument was operating within the Quality Control Value?		
10	Were the required number of readings collected for each surface?		
11	Were substrate corrections performed (if necessary)?		
12	Were confirmatory paint-chip samples collected if XRF readings were in the inconclusive range?		
13	Was procedure used to collect the paint-chip samples described?		
14	Was the laboratory that analyzed the paint samples identified?		

Figure 9.2 Form for the review of previous lead-based paint inspections.

Example of a Lead Hazard Control Policy Statement

XYZ Property Management Company is committed to controlling lead-based paint hazards in all its dwellings. _____(name), _____(position or job title) has my authority to direct all activities associated with lead hazard control, including directing training, issuing special work orders, informing residents, responding to cases of children with elevated blood-lead levels, correcting lead-based paint hazards on an emergency repair basis, and any other efforts that may be appropriate. The company's plan to control such hazards is detailed in a risk assessment report and lead hazard control plan.

(Signed) _____ _____ (Date)
(Owner)

(Signed) _____ _____ (Date)
(Lead Hazard Control Program Manager)

Figure 9.3 Example of a lead hazard control policy statement.

Condition	Yes	No
Roof missing parts of surfaces (tiles, boards, shakes, etc.)		
Roof has holes or large cracks		
Gutters or downspouts broken		
Chimney masonry cracked, bricks loose or missing, and/or obviously out of plumb		
Exterior or interior walls have obvious large cracks or holes, requiring more than routine pointing (if masonry) or painting		
Exterior siding has missing boards or shingles		
Water stains on interior walls or ceilings		
Plaster walls or ceilings deteriorated		
Two or more windows or doors broken, missing, or boarded up		
Porch or steps have major elements broken, missing, or boarded up		
Foundation has major cracks, missing material, the structure leans or is visibly unsound		
Total number*		

* If "Yes" column has two or more checks, the dwelling is usually considered to be in poor condition for the purposes of a risk assessment. However, specific conditions and extenuating circumstances should be considered before determining the final condition of the dwelling and the appropriateness of a lead-hazard screen.

Notes:

Figure 9.4 Building condition form.

Encapsulant Patch Test Documentation

Name of Person Performing Patch Test _____

License or Certificate Number (If Applicable) _____

Complete Address of Dwelling _____

Date Patch Test Applied _____ Curing Time _____

Date of Patch Test Evaluation _____

Temperature During Application and Curing _____

Humidity During Application and Curing _____

Room	Surface Location	Substrate	Type of Patch Test (X-cut or Adhesive Wallboard)	Surface Preparation	Name and Formula-tion of Encapsulant	Obser-vations	Pass/ Fail

Figure 9.5 Encapsulant patch test documentation.

Index

Note: Page numbers followed by *b* indicate boxes, *f* indicate figures, and *t* indicate tables.

Printed in the United States
By Bookmasters